Studies in Systems, Decision and Control

Volume 281

The series "Studies in Systems, Decision and Control" (SSDC) covers both new developments and advances, as well as the state of the art, in the various areas of broadly perceived systems, decision making and control–quickly, up to date and with a high quality. The intent is to cover the theory, applications, and perspectives on the state of the art and future developments relevant to systems, decision making, control, complex processes and related areas, as embedded in the fields of engineering, computer science, physics, economics, social and life sciences, as well as the paradigms and methodologies behind them. The series contains monographs, textbooks, lecture notes and edited volumes in systems, decision making and control spanning the areas of Cyber-Physical Systems, Autonomous Systems, Sensor Networks, Control Systems, Energy Systems, Automotive Systems, Biological Systems, Vehicular Networking and Connected Vehicles, Aerospace Systems, Automation, Manufacturing, Smart Grids, Nonlinear Systems, Power Systems, Robotics, Social Systems, Economic Systems and other. Of particular value to both the contributors and the readership are the short publication timeframe and the world-wide distribution and exposure which enable both a wide and rapid dissemination of research output.

** Indexing: The books of this series are submitted to ISI, SCOPUS, DBLP, Ulrichs, MathSciNet, Current Mathematical Publications, Mathematical Reviews, Zentralblatt Math: MetaPress and Springerlink.

More information about this series at http://www.springer.com/series/13304

Vitalii P. Babak · Serhii V. Babak ·
Mykhailo V. Myslovych · Artur O. Zaporozhets ·
Valeriy M. Zvaritch

Diagnostic Systems For Energy Equipments

Vitalii P. Babak
Institute of Engineering
Thermophysics of NAS of Ukraine
Kyiv, Ukraine

Serhii V. Babak
Committee on Education, Science and
Innovation of Verkhovna Rada of Ukraine
Kyiv, Ukraine

Mykhailo V. Myslovych
Department of Theoretical Electrical
Engineering
Institute of Electrodynamics
of NAS of Ukraine
Kyiv, Ukraine

Artur O. Zaporozhets
Department of Monitoring and Optimization
of Thermophysical Processes
Institute of Engineering Thermophysics
of NAS of Ukraine
Kyiv, Ukraine

Valeriy M. Zvaritch
Department of Theoretical Electrical
Engineering
Institute of Electrodynamics
of NAS of Ukraine
Kyiv, Ukraine

ISSN 2198-4182 ISSN 2198-4190 (electronic)
Studies in Systems, Decision and Control
ISBN 978-3-030-44445-7 ISBN 978-3-030-44443-3 (eBook)
https://doi.org/10.1007/978-3-030-44443-3

This Springer imprint is published by the registered company Springer Nature Switzerland AG
The registered company address is: Gewerbestrasse 11, 6330 Cham, Switzerland

Introduction

To date, from 70 to 90%, according to various estimates, of main and auxiliary equipment of the energy complex of Ukraine has developed its own resource. In these conditions, further operation of energy-intensive, and in some cases extremely dangerous (for example, nuclear power plants) equipment, requires the creation of special, scientific-based methods and means that allow such operation, ensuring the necessary level of reliability and safety. Recently, due to new information technologies and Internet, a sufficient number of such methods and tools appeared. Among them, the most effective methods are non-destructive control, monitoring and diagnostics of energy equipment (EE) units. In all these methods, the carrier of information about technical condition of studied object is a diagnostic signal. Through comprehensive study of diagnostic signal (taking into account its measurement, conversion, processing and analysis), the researcher obtains necessary information about studied object.

For a confident solution of these problems, the researcher firstly needs a mathematical model of diagnostic signal that, based on the physical features of its formation in diagnosed object, allows to obtain objective diagnostic information about this object. Secondly, he should use methods and corresponding algorithms for diagnostic signals processing that could be implemented on a basis of modern electronic devices and information technologies. Herewith, chronologically, the model is primary and the choice of methods and technical means for diagnostic signals measuring and processing is secondary.

The construction of diagnostic signal mathematical model is specified by particular physical process selected by researcher as diagnostic information carrier, and also by EE unit as an object of diagnosis. In addition, the nature of diagnostic signal model depends on chosen type of diagnosis—test or functional.

It is known that each unit of diagnosed EE has its own peculiarities in diagnostic signals formation. That is why the monograph pays considerable attention to the development and analysis of mathematical models of these signals that are based on corresponding physical processes occurring at EE units.

While constructing these models, two basic approaches could be used: deterministic and statistical. With the first approach, deterministic function of time is chosen as initial mathematical model of signal coming from primary sensors. In this case, as the most informative, amplitude-frequency and phase-frequency parameters of diagnostic signal that characterize the operation of studied EE unit are commonly used. Diagnosis by deterministic methods is basically reduced to a theoretical definition of possible diagnostic attributes and their comparison with the results of experimental data analysis. If the latter substantially differ from theoretically obtained results, then a conclusion is made about the presence of defect in studied unit. In its essence, these methods are applicable in a case when the results of all observations with the same initial conditions are identical, and also if the same parameter (characteristic) is measured under the same conditions. This situation is either highly idealized or is observed with very low accuracy of measuring instruments used, when random effects on measurements are not perceived by these instruments. Consequently, with a deterministic approach, there is no need for multiple measurements, since they are all the same and conclusion about unit technical state could be made basing on one measurement. However, there is always a possibility, by increasing the accuracy of each individual measurement, to detect the unrepeatability of measurement results in the situation described above. Then the question arises, which of the resulting series of numbers should be considered as a measurement result. Therefore, the use of deterministic methods could not be considered as satisfactory and reasoned, since many physical processes (diagnostic signals) arising in different EE units are random by their nature, i.e., their realizations vary from observation to observation. Thus, it is impossible to obtain from one observation a practically reliable answer about the technical state of diagnosed units.

Statistical approach allows us to recommend an algorithm based on a series of results of certain measurement experiment, by which the best approximation of the measured parameter to its true value is calculated in a probabilistic sense. Besides, when using statistical diagnostic methods, the measure of possible incorrect decisions about technical condition of diagnosed EE units is taken into account, and it is also possible to estimate the average number of possible incorrect conclusions, their dispersion, etc.

In most works, appropriate attention is not paid to the construction of mathematical and physical models of reference diagnostic signals, and without this, in our opinion, comparative analysis of various statistical methods of control and diagnostics is impossible. In a discussion of possibilities of statistical approach for monitoring and diagnosis, the very initial moment of diagnosis is often absent—measurements at diagnostic objects. These questions are also reflected in this monograph.

Summing up the results of a brief comparison of deterministic and statistical approaches to the construction of models, methods and diagnostic systems, the authors use the statistical approach in this study, because the vast majority of physical processes occurring in studied EE units are random by nature. Application of statistical methods for EE units diagnostic is also required by number of reasons associated with the presence of strong electromagnetic, thermal, acoustic and other fields in operating EE that act as a noise in measurement, converting, processing and analyzing information diagnostic signal.

An important point in constructing of diagnostic signal mathematical model, and then a diagnostic Information-Measurement System (IMS), based on this model, is diagnostic signal type used studying various objects. In this work, when diagnosing EE units, various types of diagnostic signals were used: vibration (vibration acceleration), acoustic emission, etc.

Working on development of IMS for EE units' diagnostics, authors of this monograph comprehensively considered problems of constructing mathematical probabilistic models of diagnostic signals, development of statistical methods for their analysis with the purpose of making a diagnostic decision and, finally, technical implementation of proposed diagnostic methods. Following the concept of primary nature of diagnostic signal mathematical model, authors found it expedient foremost to consider questions connected with the theory of random processes with infinitely divisible distribution laws, linear and linear periodic random processes. Considerable attention is paid to the problems of imitation modeling of diagnostic signals and their statistical estimation. The modern element base and new information technologies allowed authors to develop, build and practically test a number of experimental samples of information-measuring systems for statistical diagnostics of energy facilities.

A large volume of conducted experimental studies showed the operability and efficiency of constructed IMS samples.

The authors do not pretend to comprehensively consider issues on EE diagnostics using statistical methods and IMS, implemented on their basis. At the same time, the results of the studies described in this monograph are a natural continuation of the subject of statistical methods application in the field of control, monitoring and diagnostics for electric power facilities.

Contents

Chapter 1
Principles of Construction of Systems for Diagnosing the Energy Equipment

The diagnostic systems (DS) should be oriented to measuring and processing signals generated in the power equipment (PE) [1, 2]. At the same time, the principles for constructing such systems for measuring various diagnostic signals (vibration, acoustic, acoustoemission, thermal, electrical, etc.) are common. These DS include the appropriate measuring and recording equipment, as well as the necessary computer facilities and special software (software) [2, 3].

One of the most significant moments in the construction of DS is to provide a diagnostic method associated with the formation of the training sets and constructing decision rules for diagnostike and classification of certain types of defects in the test PE. These methods together with the above special software constitute the basis of information support [3, 4], which is an integral component of the product of a significant amount, constituting a so-called soft-component systems studied.

Let us dwell on these questions in more detail.

1.1 Objects of Energy Supply and Operational Reliability of Their Components

The problem of energy supply and efficient use of energy resources is one of the priority issues of Ukraine's national security. The solution of this problem directly depends on the efficiency of the main power equipment.

The main *power equipment* is defined as equipment designed to generate (electricity, steam, hot water), convert (the chemical energy of the fuel burned into steam or hot water), transport or transfer the mechanical energy of the energy carrier (water, gas, Steam, compressed air, oxygen, nitrogen, etc.).

The basic power equipment is conventionally divided into:

V. P. Babak et al., *Diagnostic Systems For Energy Equipments*, Studies in Systems, Decision and Control 281, https://doi.org/10.1007/978-3-030-44443-3_1

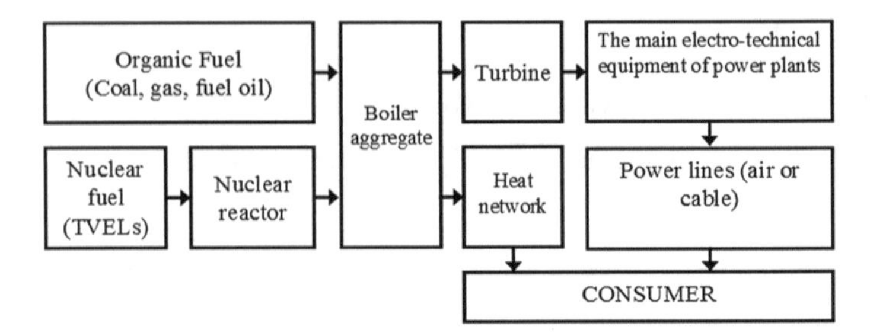

Fig. 1.1 The general scheme of the technological process of production, distribution and consumption of electric and thermal energy

Electric power [1]: generators; engines; transformers; synchronous expansion joints; switching equipment; power lines and other network equipment; means of protection and automation control; computer facilities;

Thermomechanical [5]: steam and water-heating boilers; boilers for recycling (boilers coolers); steam and gas turbines; auxiliary equipment of boiler plants; air separation units; refrigerating installations; the equipment of gas-distributing stations; compressors—centrifugal and piston; superchargers (blowers, gas blowers and exhausers), coke blowers; smoke exhausters; pumps; vessels working under pressure (energy); water pipelines (drinking, hot, technical, circulation, slimes, dewatering), gas (natural, blast, coke, etc.), steam, heat, air, oxygen, nitrogen, hydrogen and other media; channels of storm, technological, sewage; masts and supports, power lines; fittings (shut-off, regulating), service areas of pipeline fittings at elevations.

The technological process of production, distribution and consumption of electrical and thermal energy in the form of a generalized scheme is shown in Fig. 1.1.

In accordance with the above scheme, in this paper we will consider issues related to the construction of information support for systems for diagnosing energy objects.

1.1.1 Main Types of Electric Power Equipment

The basis of the electric power industry of Ukraine is the United Electric Power System (UES), which provides centralized power supply to domestic consumers. In addition, the UES cooperates with the energy systems of neighboring countries, provides import and export of electricity. It consists of eight regional electro-energy systems, interconnected by power lines up to 750 kV.

As it was mentioned above, the main electric power equipment (EPE) includes: generators, motors, transformers, synchronous compensators, switching equipment, power lines and other network equipment, control means for protection and automation, computer equipment.

Various methods for diagnosing EPE are known and are practically used [1, 2, 4, 6, 7]. These methods are mainly determined by the measured physical process, which is used as a diagnostic signal to obtain information about the technical state of the EPE node under investigation.

Technical diagnostics of power facilities are usually carried out by methods of nondestructive testing, after which these objects can be used for their intended purpose. Nondestructive testing, depending on the physical phenomena underlying it, is divided into species (DSTU2865-94): magnetic; electric; eddy current; radio wave; thermal; optic; radiation; acoustic; penetrating substances.

In this paper, as an example of obtaining diagnostic information, vibration and acousto-emission diagnostic signals appearing at the nodes of the most typical representative of EPE, namely, electric machines (EM), are considered. These signals are measured both directly on the working equipment (functional diagnostics), and on equipment that is in an inoperative state. In the latter case, the diagnosis is carried out with the help of special, most often shock, effects (test diagnostics).

To obtain information on the technical condition of the EM nodes, numerical parameters or functional characteristics of the diagnostic signal, measured directly on the EM node under investigation, are most often used. To diagnose specific EM nodes, the following parameters are usually used [6]:

- winding: electrical resistance interturn isolation; surface temperature of the windings; the magnitude of the magnetic induction;
- laminated magnetic circuit: vibration (displacement, speed, acceleration) of the frontal parts; the magnitude of the magnetic induction; surface temperature of the magnetic circuit;
- bearing assembly: vibration (displacement, speed, acceleration) of the bearing shield; temperature of the bearing shield;
- frame and the place of its attachment to the foundation: acoustic emission; vibration of the frame;
- brush-collector unit: the transient resistance (conductivity) of the sliding contact; vibration (displacement, speed, acceleration) brush holder; temperature of brush holder;
- elements of the cooling system EM: aerodynamic noise of the fan's wing fan; vibration of the surfaces of the circulation channels for moving the EM refrigerant.

Next, let us turn to the main components of the heat power equipment and present them as diagnostic objects.

1.1.2 Main Components of Thermal Power Facilities

Heat supply systems. Heat power engineering is a division of energy related to the production, use and conversion of heat into various types of energy (GOST 19431–84). There are [8] two fundamentally different directions of the use of heat—energy and technology.

With energy use, heat is converted into mechanical work, with the help of which electric energy is created in generators, convenient for transmission over a distance.

With technological (direct) use, heat is used to direct the properties of various bodies (melting, solidification, structure change, mechanical, physical, chemical properties).

The basis of modern thermal power engineering are thermal power stations [9]. Thermal power plant (TPP) is a set of interrelated installations, the general technological purpose of which is the conversion of the chemical energy of fuel into electrical energy or into electrical energy and heat (GOST 19431–84).

The main elements of the TPP are a boiler plant, a steam or gas turbine and an electricity generator. In our time steam turbine thermal power plants are predominant.

There are two types of steam turbine thermal stations—condensing power plants (CPP), designed for the production of electrical energy, and heating stations, or combined heat and power (CHP), in which the combined production of electricity and heat.

The production of heat is also possible at nuclear power plants (NPPs), in which the fission energy of atomic nuclei is converted into electrical energy or into electrical energy and heat. Steam at nuclear power plants can be obtained both in the reactor itself (single-loop NPPs) and in the steam generator (two-circuit and three-circuit NPPs).

Provision of consumers with heat is carried out by the heat supply system. According to GOST 19431-84, the heat supply system is a set of interconnected power plants that heat the district, city, or enterprise.

The main elements of the heat supply system are the source of heat energy, heat network, user input and local systems of heat consumers [3].

Let's briefly dwell on the description of the main heat power plants and some of their components.

Boiler installations. The main source of heat production is a boiler plant, which is a device for producing steam or hot water [3]. The boiler plant consists of one or more boilers and auxiliary equipment.

A boiler is a structurally integrated set of devices for producing steam or for heating water under pressure due to thermal energy from fuel combustion, the flow of a process or the conversion of electrical energy into thermal energy (GOST 23172–78).

The main elements of the boiler are [1]: furnace, surface heat exchanger, superheater, economizer, air heater. In the hot water boiler, there are no heat exchange surfaces, an air heater and an economizer.

The main elements of auxiliary equipment of boiler plants are: fuel supply system, draft equipment, water treatment devices, nutritional devices, feeding pipelines, steam pipelines, pipeline fittings, slag and ash dispenser, ash handling devices, thermal monitoring devices, boiler control units.

Thermal engines. Thermal engines are designed to convert the chemical energy of fuel into the mechanical energy of a rotating shaft [1].

A steam-turbine plant is part of a thermal or nuclear power plant. It is designed primarily for economical conversion of steam energy into the work required to drive an electric generator.

The totality of the mechanisms, apparatus and communications of the steam-turbine plant, along which its working body passes, is called the steam-water path. It includes a steam turbine, a condensation unit, a system of regenerative feedwater heating.

De-aeration-nutrient installation is formally element of a TPP or a NPPs. However, at unit power plants, each deaeration unit only caters for one steam turbine plant and technologically is an integral part of it. The system of intermediate dehumidification (separation) and steam-steam overheating also belong to the steam-water path of NPPs and TPP.

A gas turbine plant is a heat engine, the working body in which remains gaseous at all points of the thermal cycle; consists of turbines, compressors, intake devices (combustion chambers) and heat removal combined with a common hydromechanical system.

Depending on the method of transferring part of the heat to the cold source, gas turbine installations of open and closed cycles are distinguished. The working body of an open-cycle gas turbine plant is atmospheric air and combustion products of organic fuel, and in closed gas turbine installations—air, helium, nitrogen, carbon dioxide, etc.

In an open-cycle gas turbine plant, the working fluid comes from the atmosphere, passes through all the elements of the unit once and emits into the atmosphere. In a gas-turbine installation of a closed cycle, the working medium continuously circulates through a closed circuit, and heat removal takes place in special heat exchangers.

The gas piston engine is an internal combustion engine in which gaseous fuel is used. The gas piston engine consists of a casing, the main element of which is a cylinder, as well as a crank mechanism, a gas distribution mechanism, a gas supply system, an air inlet and exhaust system, lubrication systems, engine cooling systems, ignition and start systems.

Cogeneration plants are designed for combined production of electrical and thermal energy. The main nodes of the basic scheme of the cogeneration unit are [1] a thermal engine, an electric generator and a heat recovery unit.

Electric energy is produced by a thermal engine and an electric generator, thermal energy is a thermal engine and a heat-exchanger. The cogeneration unit also includes auxiliary equipment—draft machines, pipelines and control systems. The main parameters of the cogeneration plant are the thermal and electric power of the installation, the efficiency in the generation of electricity, the fuel utilization factor, the temperature of the exhaust gases and their number at the rated power [10]. Let us dwell on a brief description of the main components of the cogeneration plant.

The thermal engines of the cogeneration unit convert the chemical energy of the fuel into mechanical energy of the rotating shaft. The exhaust gases of a thermal engine are used to generate heat energy.

Electric generators convert the mechanical energy of the rotating shaft of the heat engine into electricity.

The heat exchanger serves for the generation of thermal energy by using the energy of the exhaust hot gases from the heat engine.

Heating network. The transfer of heat from the source to the consumers is carried out with the help of heat networks. According to DBN B.2.5–39: 2008, the following heat networks are distinguished:

- Main heat network—a complex of pipelines (pipelines) and facilities that provide transportation of heat carrier from a source of thermal energy to heat points and (or) a distribution heat network.
- Distribution heat network—pipelines with structures on them that provide transportation of the heat carrier from the central heating station, or the main heat network to the thermal input of the consumer.
- Hot water supply network—a complex of pipelines (heating pipes), equipment and facilities that provide the supply of hot water from a heat point or from a source of thermal energy to the hot water supply of a consumer.

Subscriber installations perform docking of heat networks of district heating systems with local systems of heat consumption.

The composition of the subscriber unit is determined by the schemes for connecting the heating and hot water supply systems, the parameters of the heat carrier, the modes of heat consumption,

Subscriber installations are equipped with hot water heaters, elevators, pumps, fittings, control and measuring devices for regulating parameters and flow of coolant through local heating and water distribution devices.

1.1.3 Main Indicators of Operational Reliability of Energy Facilities

Some indicators of reliability of electrical power equipment. As of 2015, the technical condition of the electric power industry is unsatisfactory [1]. It is necessary to modernize and introduce new resource-saving technologies, to develop alternative sources of electricity (solar, wind, biogas and geothermal power plants). It is this fact that makes it necessary to carry out works aimed at ensuring the reliability of EE. One of the directions of the solution of this problem is the creation of new methods and means of monitoring and diagnostics of the specified equipment.

As noted above, the main EE of power objects is: generators, various types of electric machines (EM), used as auxiliary engines, synchronous compensators, transformers (power, measuring, etc.), switching equipment (switches, disconnectors and so on), network hardware.

The averaged reliability indicators of the main EE [1], given in Table 1.1, indicate that the most unreliable EO can be attributed to generators, electric motors, transformers and their nodes.

This is confirmed by the data indicated in the table: synchronous generator (failure rate = 1, average recovery time = 100), asynchronous motor (failure rate = 0.1, average recovery time for motors LV = 50, and for motors HV = 160). Therefore,

Table 1.1 Averaged reliability indicators of the main EE

Type of electrical equipment	$\lambda, 1/Y$	T_B, h
Disconnector	0.01	2
Short-circuiting device	0.02	10
Separator	0.03	10
Low voltage circuit breaker (LV)	0.05	4
High voltage fuse (HV)	0.1	2
Busbars with voltage up to 10 kV (per one connection)	0.03	2
Cable line VN, laid in:		
Trench	0.03	44
Blocks	0.005	18
HV cable line laid in the trench (per 1 km)	0.1	24
The air line of the NN (1 km)	0.02	5
Synchronous generator	1	100
Asynchronous motor:		
HV	0.1	50
LV	0.1	160

in this paper, we will focus mainly on the diagnosis of these types of equipment. The methods and tools considered in this work are oriented to the specified equipment, but can also be used for other units and devices operating on similar principles of electrical energy conversion.

As the operating experience shows, the main cause of EM accidents and repairs is insulation failure both between windings and between the winding and the hull. According to [1], the second after the winding by the number of failures is the bearing assembly. In the vast majority of cases, failures of generators and electric motors occur due to damage to the windings (85–95%). From 2 to 5% EM is rejected because of bearing damage. The remaining nodes (brush-collector unit, ventilation system, etc.) account for 1–2% of failures.

To obtain information about the technical state of the nodes of electrotechnical equipment, various processes that arise in it during its operation are used, namely their numerical parameters and functional characteristics. To diagnose specific EM nodes, the following parameters are usually used:

- winding: electrical resistance interturn isolation; surface temperature of the windings; the magnitude of the magnetic induction;
- laminated magnetic circuit: vibration (displacement, speed, acceleration) of the frontal parts; The magnitude of the magnetic induction; surface temperature of the magnetic circuit;
- bearing assembly: vibration (displacement, speed, acceleration) of the bearing shield; temperature of the bearing shield;

- frame and the place of its attachment to the foundation: acoustic emission; vibration of the frame;
- brush-collector unit: the transient resistance (conductivity) of the sliding contact; vibration (displacement, speed, acceleration) brush holder; temperature of brush holder;
- elements of the cooling system EM: aerodynamic noise of the fan's wing fan; vibration of the surfaces of the circulation channels for moving the EM refrigerant.

Let us briefly discuss some issues related to the primary research (measurement) of the above processes, which, when solving the problem of diagnosing EM, act as diagnostic signals.

Windings EM. Reliability of EM is largely determined by the reliability of their windings, which, in turn, depends on the state of their insulation. The latter operates under conditions of strong electromagnetic, thermal, and vibrational fields. In addition, the external climatic conditions, in which EM is operated, have a significant effect on the insulation state. Over time, these conditions together lead to a progressive deterioration in the properties of insulation. It should be noted at once that a large number of works have been devoted to the control and diagnostics of isolation, for example, [11].

The main characteristic of insulation, which determines the reliability of EM operation, is its electrical strength. However, this most important property of insulation can be maintained in the process of exploitation only in the presence of many other qualities, the lowering of which leads to a decrease in electrical strength. Thus, insulation must maintain a high thermal conductivity, otherwise there will inevitably be an increase in local heating, accompanied by accelerated destruction. The insulation must have sufficient mechanical strength and elasticity, which would exclude the possibility of formation of residual deformation, cracks, stratification of it under the influence of mechanical forces. Isolation should maintain a stable chemical composition, since its change leads to a decrease in its electrical strength.

In high-voltage machines, the aging of insulation under the influence of an electric field is of great importance. During operation, the EM insulation for a long time is under operating voltage and periodically is affected by increased stresses—with preventive tests and various wave phenomena, the sources of which can be located both outside and inside the machine.

The operation data and experimental studies [1] show that a noticeable effect of the electric field on the service life of insulation begins to be detected in machines with a rated voltage of at least 6 kV. In machines for lower stresses, the phenomena of electrical aging are not observed.

The insulation of the slotted part of the winding can undergo compression under the influence of electrodynamic forces, and in the presence of gaps in the groove it is also subject to impacts and abrasion against the groove walls. If there is no freedom of movement in the groove, in isolation, in addition to tension and compression, shear deformation is also possible. When bending the frontal parts of the winding, the greatest stresses occur at the exit points of the rods or coils from the grooves, where the insulation experiences compression and tension stresses. In addition, it crumples on the pads and in contact with the bandages.

In most cases, these efforts have a cyclic, alternating character, with the most typical vibration being the frequency of 100 Hz [1, 8, 9]. Periodically, during transient processes (starting and reverse of motors, short circuits and non-synchronous switch-on of generators), the vibration amplitudes increase tens of times, due to the increase in currents in the windings and the quadratic dependence of the electrodynamic forces on the current. Particularly significant efforts can occur in the windings of large machines, turbo and hydro generators.

The mechanical characteristics of insulation depend on its temperature. As the insulation is heated, the insulation strength decreases rapidly, and at the same time the insulation becomes more elastic. This is especially true for insulating structures on thermoplastic compasses. Thus, the tensile strength of the mica compound under tension is 3340 N/cm^2 (340 kgf/cm^2) at 20 °C and only 344 N/cm^2 (35 kgf/cm^2) at 100 °C. These values vary greatly depending on the features of technology, operating time and other reasons; They are also different for different parts of the winding [1]. With a decrease in temperature, this insulation becomes fragile. The mechanical characteristics of insulation on thermosetting binders are more stable.

According to [1], the destruction of insulation occurs gradually, and the initiating role belongs to the processes of thermal aging. Even at relatively low temperatures, when the thermal-oxidative destruction is negligible, the drying of insulation, the evaporation of volatile components from the binders, the decrease in the elasticity of the insulation, and the increase in its brittleness occur under the action of heat. The latter contributes to the development of mechanical aging processes. Cracks and other defects appear in the insulation, it breaks up and loosens, which creates the conditions for the occurrence of ionization phenomena. The destruction of insulation occurs unevenly and ends with a breakdown in the weakest place. Moisture and aggressive environments contribute to the acceleration and activation of aging processes.

As a basic diagnostic feature, which makes it possible to judge the electrical strength of insulation, the value of the resistance of the interturn isolation is usually used.

At present, a large number of devices, devices and systems for determining the electrical strength of insulation can be found on the market [11].

Fused magnetic circuit. The technical condition of the windings of the EM stator is directly related to the state of the laminated magnetic wire. To date, for stranded magnetic cores of a stator of powerful EM, one of the most common faults is the defects associated with weakening the pressing of the extreme packets of the stator's frontal parts (the so-called "swelling"), as well as the mechanical breakage of the tightening rods or nuts that fasten the outer clamping flange. The results of recent studies [12, 13] show that these defects are often associated with the manifestation of "anomalous vibromechanical phenomena" in the stator cores of powerful EM. Especially these phenomena become noticeable when, with an increase in the diameter of the stator core, the EM's own frequency begins to approach the frequency of the radial forces of magnetic tensions in the air gap.

It is possible to effectively detect such defects by applying statistical spectral analysis to the processing of vibration diagnostic signals (vibration displacement, vibration velocity and vibration acceleration), which are measured with the help of appropriate vibration sensors. Measurement, recording and subsequent processing of these vibrations can be carried out using automated DS vibration diagnostics. A description of some of them is given, for example, in [14, 15]. In the implementation of these experiments, in the case of the use of these DS vibrodiagnostics, the primary sensors are located in the end parts, on the pressure flanges of the stator (or drum rotor) of the EM. This approach [1] also makes it possible to determine the degree of compacting of the stranded magnetic core of the stator (and for some types of EM and rotor) of EM.

Additional information on the technical state of the EM core can also be obtained from the temperature data measured with different thermocouples located directly in the grooves of the core of the EM magnetic core.

Bearing assembly. The most informative physical process, which makes it possible to diagnose almost all types of defects in rolling bearings, is vibration (vibration displacement, vibration-speed, vibration acceleration) [14, 15]. Following the requirements of GOST 12379-75, the corresponding sensors are placed on the bearing units of the EM in three mutually orthogonal directions (X, Y, Z axes). In the X and Y axes, the vibrations are measured in the radial direction (relative to the rotating shaft EM), and along the Z axis in the axial direction. Synchronous vibration measurement in these three directions allows using the appropriate processing to explore the vibration field, which greatly expands the diagnostic capabilities.

The high information content of the vibration diagnostic signal measured at the EM bearing unit is due to the fact that the rolling bearings are the only contact connection between the stationary and rotating parts of the EM and in them the vibrations perturbed at different nodes of the machines under investigation are concentrated and mutually superimposed.

The base and the places of its fastening to the foundation. Some general information on the technical condition of EM can be obtained from the diagnosis of the EM frame and the places of its attachment to the foundation.

Among the processes that make it possible to judge the general state and mechanical loading of massive assemblies of EM attachment are the processes of acoustic emission (AE) [16, 17]. To measure the characteristics of AE processes, the relevant sensors must be located directly on the diagnostic nodes. A distinctive feature of the processes of AE is their rather wide frequency range, the upper limiting frequency of which for some materials reaches several megahertz.

Along with AE processes, additional information on the technical state of EM during operation can be obtained by measuring the vibration process. In this case, according to DSTU ISO 5348: 2009, vibration sensors are placed on the bearing supports (for high-power EM), and also directly at the EM attachment points to the foundation.

Brush-collector unit. In operation of most powerful EMs that have an excitation system based on current transmission to the rotating field winding through sliding contacts, special attention is required to the operation of the brush-collector unit.

Nevertheless, as noted in [1], until now the creation of technical diagnostic tools both in Ukraine and abroad is constrained, first of all, by the absence of theoretically grounded and experimentally verified diagnostic features and criteria that would allow with a given reliability to carry out diagnostics of the brush-collector unit [1].

The most successful attempt to theoretically substantiate and test experimental diagnostic features that determine the technical state of the brush-collector device was made in [1]. In this work, it is recommended to use the following 5 physical processes to monitor and diagnose this device:

sparking of brushes;

- the radio emission of the brush and its magnitude;
- voltage drop ΔU on the brush transition—ring;
- ambient air temperature near the brush unit;
- uneven distribution of current on individual brushes in the brushing device EM.

One of the main parameters that is used to monitor the state of the brush-collector unit is the transient resistance (conductivity) of the sliding contact. This parameter is measured using special sensors, which are installed directly on the brush holders.

Elements of the cooling system of EM. Diagnosis of the EM ventilation system can be realized by measuring the acoustic noise and vibrations of its individual components (airway surface, fan fan wing, fan shaft bearings).

To measure the acoustic noise of the EM ventilation system, a wide range of acoustic measuring equipment can be used, for example, the Brüel & Kjær firms of type 8609, 8637, 8752, etc. Practical measurements of acoustic noise EM are carried out taking into account the requirements of GOST 17377–73, 17975–75, 17995–75 and others.

Measurement of EM vibrations is realized with the help of primary sensors, which can be installed both on the surface of the air duct and on the bearing units of the fan. In this case, it is necessary to ensure compliance with the requirements of GOST 17375–75, 17377–73, etc.

1.2 Generalized Structure of Systems for Diagnosing Energy Objects

Information on the state of the operating equipment is contained in various signals, which are measured during its operation. One of the indicators is the temperature of various parts of the EE. For EE with rotating nodes, the most important diagnostic signal is vibration. Therefore, the analysis of vibrations and temperature of individual components is the basis for the technical diagnosis of EO. This line of research has broadened the capabilities of existing methods of nondestructive testing, enabled us to solve practical problems of long-term prognosis of the state of equipment that contains rotating nodes.

Recently, the Interstate Council for Standardization, Metrology and Certification, which includes Ukraine, adopted a number of standards related to the diagnosis of electrical equipment, and comply with international standards ISO (International Standard Organization). Let's consider some of them, which should be used when creating modern diagnostic equipment for PE.

The analysis of the current standards of Ukraine, interstate, European and international standards in the field of ensuring the reliability of PE gives grounds to formulate the following requirements, which modern competitive systems of vibration diagnostics of EE should meet.

Measuring ranges:

- vibration speed (not less than): 30 mm/s;
- vibration displacement (not less than): 350 μm.

Frequency range:

- measuring range of vibration (not worse): 10–2500 Hz;
- accelerometer calibration frequency: 79.6 Hz.

Nodes, on which it is necessary to measure vibration:

- for synchronous machines: vibration on the upper caps of the bearings in the vertical direction and at the connector in the transverse and axial directions;
- in hydrogenerators: vibration in the horizontal plane of the crosses;
- in turbogenerators: vibration of contact rings;
- for turbo and hydro generators: vibration of the core and the frontal parts of the stator winding.

Accelerometer mountings (DSTU ISO 5348: 2009) with the help of:

- hairpins;
- methyl cyanoacrylic adhesive;
- beeswax;
- double-sided adhesive tape;
- vacuum mounting;
- magnet;
- hand probe.

Accuracy of measurements—according to GOST 25275–82/ ST SEV 3173–81. The electrical control device must have an electrical calibration device that must output a harmonic signal at a frequency of 79.6 Hz; The error of electrical calibration of vibration measuring instruments should not exceed 5%; The basic error in measuring the harmonic vibration within the working part of the scale should not exceed 10%.

Taking into account the above main provisions of the standards regarding the equipment for measuring and diagnosing electrical equipment, we turn to the consideration of the general structure and especially the construction of DS diagnostics.

1.2.1 Features of Construction of Systems for Diagnosing Energy Objects

From the above list of diagnostic parameters and signals it follows that the latter have sufficiently wide dynamic and frequency ranges. With this in mind, modern AIS EM diagnostics should be universal, i.e. capable of measuring, recording, and processing diagnostic signals of a different physical nature, as well as dynamic and frequency ranges. Let us briefly dwell on these questions.

The generalized block diagram of the modern multichannel IMS of functional diagnostics is shown in Fig. 1.2. It consists of:

- measuring transducers designed to measure the various physical processes that accompany the operation of the object;
- analog part intended for preliminary processing of the measured information (coordination of resistances, amplifications, filtering, etc.);
- interface between the analog and digital parts, whose role is reduced to converting the measured signals into digital form and transferring them to the digital part of the DS;
- a digital part intended for the subsequent processing of information (for example, digital filtering) and analyzing the state of the object in accordance with a predetermined algorithm, detecting and classifying defects;
- user interface, through which the operator performs control of the operation of the system and receives information about the state of the object.

The digital part of IIS diagnostics in most cases is a personal computer that operates under the control of specialized software. In IIS diagnostics of geographically distributed objects, or if access to nodes that are diagnosed is complicated (for example, permanently rotating blades of wind power aggregates), it is possible to use special channels for transmitting information from measuring converters over considerable distances (over a radio channel, fiber-optic lines and the like).

The development of diagnostic tools for a particular type of equipment is associated with the solution of the following main tasks [18]:

- the definition of the class of possible defects (the most important or most frequently encountered) that must be identified;

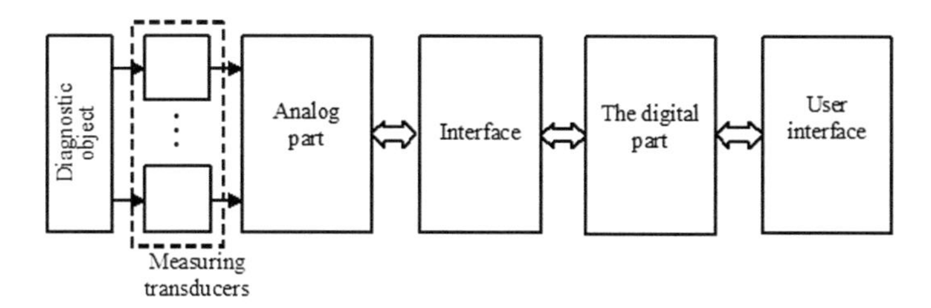

Fig. 1.2 Generalized block diagram of DS diagnostics

- selection of diagnostic signals available for measurement, and control points on the object under study;
- development of a mathematical model of the diagnostic object, the analysis of which allows to substantiate possible diagnostic parameters;
- development of algorithms for obtaining numerical values of selected diagnostic parameters;
- construction of decisive rules for identifying and classifying defects;
- creation or technical means that implement certain stages of the diagnostic process: from measuring the selected diagnostic signals to making diagnostic decisions.

For the first task, basically, statistical material on failures is required. To solve the second task, it may be necessary to make constructive changes to the diagnostic object in order to ensure the fastening of the respective measuring transducers.

The solution of the third and fourth problems are interrelated, namely, from the developed mathematical model of diagnostic signals measured at the diagnostic object, algorithms for obtaining numerical values of the selected diagnostic parameters depend.

In solving the fifth problem, the defining moment is the type of the developed mathematical model (deterministic or statistical). Depending on the chosen model, the construction of decision rules is carried out, which allow to define and classify the types of defects in the nodes of the diagnosed equipment.

The solution of the last task related to the development of the DS diagnostics proper is carried out on the basis of the results of the solution of the five problems formulated above. Basically, this applies to various types of software that are equipped with the developed DS.

1.2.2 Informational Support of Systems

The term information support requires a certain interpretation and specification, since information support is an integral product of a significant number of components that make up the so-called soft component of diagnostic systems. For example [3, 5], information support in information systems is a set of forms of documents, normative databases and implemented solutions for the volumes, location and forms of information existence. To this kind of information security, the requirements of integrity, completeness, reliability, protection against unauthorized access, unification, minimization of the volume in the transmission and preservation are put forward.

If we talk about the work of modern hardware and software systems, then from the point of view of reliability, the hardware of systems is aging means, since the action of such processes as aging, changing characteristics, parameters, elements in time, etc. significantly limit the time interval of their work. Information support is referred to as ageless means, therefore, in the process of working such software, primarily software, reduces the number of errors and has a significant time period of application. In turn, increasing the load on information support when performing the specified monitoring, monitoring and diagnostic functions makes it possible to increase the reliability of such systems as a whole.

It is well known that, on the other hand, the potential capabilities of modern computer facilities can be realized by various software variants that can largely provide the solution of various research problems without changing the hardware of computer systems and, thereby, save money and time to develop new and modernize existing ones systems. In general, information support is an important information resource for the development of various objects, systems, complexes.

The integrated level of information support of technical systems is determined at each stage of the development of science and technology (dozens of years), the development of the scientific and methodological basis, primarily theoretical, including the mathematical apparatus.

Along with servicing existing monitoring systems, information support should serve as a basis for creating a scientific and technical basis for the development of prospective systems. If one tries to introduce a certain classification of information support of technical systems, different classification characteristics can be taken as a basis. So following signs of classification characterize both current, and perspective variants of performance of the set functions by monitoring and diagnostic systems, for example:

- functionally-current information support, which remains largely unchanged during the development, implementation and operation of the current system sample;
- innovative information support, which reflects both the results of operation of existing monitoring and diagnostic systems, as well as new developments in various fields of science and technology.

With this in mind, the following classification is quite reasonable:

- a priori support at the first stages of the life cycle of systems, namely at the stages of design, manufacture, testing, when the calculation characteristics and parameters of the system are mainly used;
- a posteriori support is mainly used in the operation stages and in part of the system tests and the content is an adjusted security, the application of which enables the calculation characteristics to be corrected and to obtain real system characteristics, including reliability characteristics, residual life by statistical estimation of real measurement data, diagnosis and prognosis.

Other signs of classification of information support in more detail reflect the monitoring and diagnostic process:

- theoretical, algorithmic and software, or scientific and technical, at all stages of the application of these systems. Particularly responsible role of its use at the initial stages, when physical and mathematical models and systems are being developed, the system structure is substantiated based on the results of computer

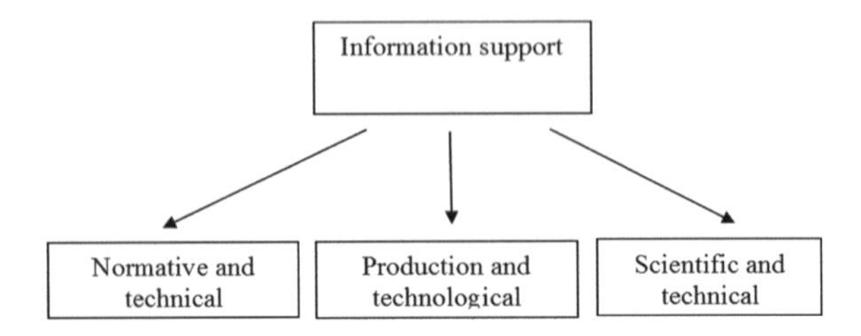

Fig. 1.3 Conditional general classification of information support

modeling, the design characteristics, parameters of the system's performance of given functions are determined;

- production and technological support at the stages of preparation of production, production of experimental and serial models of the system;
- normative and technical support is used at all stages, but the main role is played by testing, certification, transfer to operation and at the stage of operation, repair and modernization of the system.

Such a general view of information support is presented in Fig. 1.3.

The weight and importance of each of these types of information support is different at different stages of the life cycle of monitoring and diagnostic systems. The regulatory and technical and design information technology services consist of a generally accepted regulatory framework, standards, guidelines, existing databases, and sanitary norms. Their main function is to inform the existing systems that implement known, generally accepted principles and methods of diagnosis. In a sense, they can be viewed as a routine component of information support.

Scientific and technical information support is oriented to a greater extent for use in innovative developments. Such developments primarily include:

- improvement of existing systems and development of fundamentally new monitoring and diagnostic systems;
- expansion of the scope of their application and range of controlled characteristics and parameters;
- increase the effectiveness of monitoring systems based on the use of new physical effects and phenomena for constructing the converter components of systems;
- new models of information signals, their characteristics and parameters, which most adequately reflect the objective reality;
- new methods of information processing, methods for identifying and highlighting the most informative diagnostic features;
- wide application of methods of statistical analysis of experimental data;
- inclusion in the range of controlled new information characteristics;
- evaluation of reliability of hardware and software monitoring and diagnostic systems;
- forecasting the state of monitoring and diagnostic systems.

Therefore, scientific and technical information support systems for monitoring and diagnostics of energy facilities should be based on the achievement of fundamental and applied technical sciences, primarily mathematics, physics, informatics, mathematical statistics, reliability theory of technical systems, computer engineering, and the like.

The authors did not intend to cover in this work the whole range of issues related to information support for diagnostic systems. In our opinion, the most important for accelerating the development of diagnostic systems is the scientific and technical component of their information support. It is these issues that are considered in this monograph.

1.3 Physical Processes Characterizing the Technical Condition of Energy Facilities

Consider the basic physical processes (diagnostic signals) that are used in the diagnosis of EE.

First of all, it should be noted that in the operated EO there are both mobile and fixed nodes. This affects both the ways of measuring the diagnostic signals, and the methods of processing them (filtration, accounting for the cyclicity due to the frequency of the rotating electromagnetic field, etc.). This circumstance must be taken into account when choosing a mathematical model for describing the physical processes under study at the EO nodes. Here we will dwell on the main features of various diagnostic signals.

Many models are based on the fact that at the point of measurement at any fixed time the process under investigation can be considered as the result of imposing a large number of perturbing factors characterizing the energy coming from different random sources located in space at some distance from this point. Such models can be used in describing noise and vibration processes and fields, acoustic emission processes that arise in a solid under its mechanical loading or, for example, the processes of the occurrence of electrical discharges in insulation when high voltage is applied. Random nature in electrical equipment is also a process that has some stochastic cyclicity. These include the unevenness of the rotation of the shafts of diesel-electric aggregates, rolling bearings, etc.

As shown in [19], mechanical movements or vibrations are one of the most informative processes that can be used to diagnose EE. Proceeding from this, let us dwell on the appearance of perturbing forces in various nodes of EM, causing their vibration.

Technical diagnostics of energy facilities should be carried out using methods of nondestructive testing, after which the objects can be further used for their intended purpose.

A class of physical methods of non-destructive testing are methods based on the action of physical fields or substances on an object or on the registration of fields created by the object of control itself.

Nondestructive testing, depending on the physical phenomena underlying it, is divided into species (DSTU 2865–94): *magnetic; electric; eddy current; radio wave; thermal; optic; radiation; acoustic; penetrating substances.*

Within the framework of each type of non-destructive testing, there are currently a large number of different methods that are classified according to the following criteria (GOST 2865–94):

- the nature of the interaction of physical fields or substances with the object of control;
- primary informative parameter;
- way of obtaining primary information.

We give a brief description of the types of nondestructive testing and methods based on the nature of the interaction of physical fields with the object of control.

Magnetic non-destructive testing is based on the analysis of the mutual fashion of the magnetic field with the object of control. Methods of this kind do not classify by the nature of the interaction of the field with the object.

The magnetic form is usually used to control objects from ferromagnetic materials. With the help of magnetic methods, quenching and grinding cracks, hair, fatigue cracks and other surface defects with a width of several millimeters can be found.

Electrical non-destructive testing is based on the registration and analysis of the parameters of the electric field interacting with the object of control or the field arising in the object of control as a result of external influence. By the nature of the interaction of the field with the object, the methods are: *electric; triboelectric; thermoelectric.*

The electrical view is applied to the control of items made of electro-transparent materials, electrical equipment, deviations of the shape of the object.

Eddy current non-destructive testing is based on the analysis of the interaction of an external electromagnetic field with the electromagnetic field of the eddy currents induced in the object of control. By the nature of the interaction of the field with the object, the methods are:

- penetrating radiation;
- reflected radiation.

Eddy current view is used only for products made of electrically conductive materials. The peculiarity of the eddy current control is that it can be carried out without contact between the converter and the object, which makes it possible to obtain good control results even at high velocities of the objects.

Radio wave non-destructive testing is based on the registration and analysis of changes in the parameters of the electromagnetic field of the radio range, which interacts with the object of control. Usually, radio waves of wavelength 1–100 mm are used. By the nature of the interaction of the field with the object, the methods are:

- penetrating radiation;
- reflected radiation;

- scattered radiation;
- resonant.

The radio wave form is used to control products from materials in which radio waves do not strongly decay, for example, dielectrics, ferrites, thin-walled metallic objects.

Thermal non-destructive testing is based on the recording and analysis of changes in thermal or temperature fields of control objects caused by defects. By the nature of the interaction of the field with the object, the methods are:

- thermal contact;
- convective;
- own radiation.

Thermal view is used to control various objects from any materials.

Optical non-destructive testing is based on recording and analyzing the parameters of optical radiation interacting with the object of control. By the nature of the interaction of the field with the object, the methods are:

- transmitting radiation;
- reflected radiation;
- scattered radiation;
- induced radiation.

Optical control methods have low sensitivity, therefore they are used exclusively for detecting large surface defects, leak marks, residual deformations, structural damage, etc.

Radiation non-destructive testing is based on the registration and analysis of ionizing radiation after interaction with the object of control. Depending on the nature of the ionizing radiation, the radiative form is divided into subspecies, for example, X-ray, neutron, and the like. By the nature of the interaction of the field with the object, the methods are:

- penetrating radiation;
- scattered radiation;
- activation analysis;
- characteristic radiation.

The radiation view is used to control objects from various materials—large cast parts, thick-walled blanks and welded joints.

Acoustic nondestructive testing is based on recording and analyzing the parameters of elastic waves that are excited or appearing in the object of control. When using elastic waves of the ultrasonic range (over 20 kHz), the term “ultrasonic” is usually used instead of the term “acoustic”. By the nature of the interaction of the field with the object, the methods are: *penetration; reflections; resonant; impedance; free vibrations; acoustic emission.*

Acoustic methods are widely used in the control of welded joints, hulls, pressure vessels, large cast parts, thick-walled billets, etc.

Non-destructive testing by penetrating substances is based on the penetration of test substances (liquid or gas) in the cavity of defects in the object of control. The methods of this species are divided into two subspecies—capillary and percolation.

Methods of leakage are used only to detect end-to-end defects, capillary methods to identify invisible or poorly visible surface and through defects in control objects.

In the process of operating power equipment, the main role is played by the systems of functional diagnosis, the source of information in which are noise and rhythmic signals arising from the natural functioning of objects.

Noise signals are a consequence of aerodynamic, hydrodynamic and tribomechanical processes accompanying the operation of power equipment units, and are manifested in the form of acoustic, magnetic, electric, thermal noise or broadband vibrations.

Rhythmic signals are the result of interaction of parts in the kinematic pairs of gas turbines, gas piston engines, electric machines, compressors, etc., and are manifested, as a rule, in the form of narrowband multifrequency vibrations.

In Table 1.2 shows the sources and types of noise and rhythmic information signals in power equipment.

It should be noted that the systems of functional diagnostics, based on the use of noise and rhythm signals, are widely used to determine the technical state of power equipment [1, 2, 6].

Among such systems, the most widely used acoustic emission systems [1, 2, 6] diagnoses, which are based on the use of noise signals. In turn, models of rhythmic signals are used in the construction of vibrodiagnostic systems [8, 16, 19].

Table 1.2 Sources and types of noise and rhythm signals

No	Signals sources	Types signals
1	Boilers	Temperature fluctuation
2	Gas turbines	Aerodynamic noise and vibration, friction-sliding noise, acoustic emission signals
3	Steam turbines	
4	Gas piston engines	
5	Electric generators	Aerodynamic noise and vibration, friction-sliding noise, signals acoustic emission, magnetic noises, contact noise in collector-brush assemblies
6	Electric motors	
7	Piping of boiler plants	Acoustic leakage signals, acoustic emission signals
8	Pumps, compressors	Hydrodynamic noise and vibration
9	Fans	Aerodynamic noises and vibrations
10	Heating network pipelines	Hydrodynamic noise and vibration, acoustic emission signals, acoustic leakage signals
11	High-voltage equipment	Partial discharges, magnetostrictive noise in transformers

References

1. Babak, S.V., Myslovych, M.V., Sysak, R.M.: Statistical diagnostics of electrical equipment (2015). ISBN 978-966-02-7704-5
2. Babak, V.P.: Hardware-software for monitoring the objects of generation, transportation and consumption of thermal energy (2016). ISBN 978-966-02-7967-4
3. Zaporozhets, A.A., Eremenko, V.S., Serhiienko, R.V., Ivanov, S.A.: Development of an intelligent system for diagnosing the technical condition of the heat power equipment. In: IEEE 13th International Scientific and Technical Conference on Computer Sciences and Information Technologies (CSIT), pp. 48–51 (2018). https://doi.org/10.1109/stc-csit.2018.8526742
4. Zaporozhets, A., Eremenko, V., Isaenko, V., Babikova, K.: Approach for creating reference signals for detecting defects in diagnosing of composite materials. In: Shakhovska, N., Medykovskyy, M. (eds.) Advances in Intelligent Systems and Computing IV. Springer, Cham, vol. 1080, pp. 154–172 (2020). https://doi.org/10.1007/978-3-030-33695-0_12
5. Czichos, H.: Handbook of Technical Diagnostics. Fundamentals and Application to Structures and Systems (2013). ISBN 978-3-642-25850-3
6. Babak, V.P.: Information support for monitoring of thermal power facilities (2015). ISBN 978-966-02-7478-5
7. Stognii, B., Kyrylenko, O., Butkevych, O., Sopel, M.: Information support of problems of electric power systems control. Energy Econ. Technol. Ecol. **1**(30), 13–22 (2012)
8. Edwards, S., Lees, A.W., Friswell, M.I.: Fault diagnosis of rotating machinery. Shock. Vib. Dig. **1**(30), 4–13 (1998)
9. Zaporozhets, A., Eremenko, V., Serhiienko, R., Ivanov, S.: Methods and hardware for diagnosing thermal power equipment based on smart grid technology. In: Shakhovska, N., Medykovskyy, M. (eds.) Advances in Intelligent Systems and Computing III. Springer, Cham, vol. 871, pp. 476–489 (2019). https://doi.org/10.1007/978-3-030-01069-0_34
10. Zaporozhets, A.: analysis of control system of fuel combustion in boilers with oxygen sensor. Period. Polytech. Mech. Eng. **64**(4), 241–248 (2019). https://doi.org/10.3311/PPme.12572
11. Napolitano, A.: Generalizations of Cyclostationary Signal Processing: Spectral Analysis and Applications (2012). ISBN 9781119973355
12. Yatsuk, V., Mykyizhuk, M., Bubela, T.: Ensuring the measurement efficiency in dispersed measuring systems for energy objects. In: Królczyk, G., Wzorek, M., Król, A., Kochan, O., Su, J., Kacprzyk, J. (eds.) Sustainable Production: Novel Trends in Energy, Environment and Material Systems. Studies in Systems, Decision and Control, vol. 198, pp. 131–149 (2020). https://doi.org/10.1007/978-3-030-11274-5_9
13. Chen, P., Taniguchi, M., Toyota, T., He, Z.: Fault diagnosis method for machinery in unsteady operating condition by instantaneous power spectrum and genetic programming. Mech. Syst. Signal Process. **1**(19), 175–194 (2005). https://doi.org/10.1016/j.ymssp.2003.11.004
14. Brie, D.: Modelling of the spalled rolling element bearing vibration signal: an overview and some new results. Mech. Syst. Signal Process. **3**(14), 353–369 (2000). https://doi.org/10.1006/mssp.1999.1237
15. McCormick, A.C., Nandi, A.K.: Cyclostationarity in rotating machine vibrations. Mech. Syst. Signal Process. **2**(12), 225–242 (1998). https://doi.org/10.1006/mssp.1997.0148
16. Williams Jr., J.H., DeLonga, D.M., Lee, S.S.: Correlations of acoustic emission with fracture mechanics parameters in structural bridge steels during fatigue. Mat. Eval. **40**(10), 1184–1189 (1982)
17. Krishnakumari, A., Elayaperumal, A., Saravanan, M., Arvindan, C.: Fault diagnostics of spur gear using decision tree and fuzzy classifier. Int. J. Adv. Manuf. Technol. **9–12**(89), 3487–3494 (2017). https://doi.org/10.1007/s00170-016-9307-8

18. Babak, S., Babak, V., Zaporozhets, A., Sverdlova, A.: Method of statistical spline functions for solving problems of data approximation and prediction of object state. In: CEUR Workshop Proceedings, vol. 2353, pp. 810–821 (2019) (Online). http://ceur-ws.org/Vol-2353/paper64.pdf
19. Domanski, P.D.: Statistical measures. In: Control Performance Assessment: Theoretical Analyses and Industrial Practice. Studies in Systems, Decision and Control, vol. 245, pp. 53–74 (2020). https://doi.org/10.1007/978-3-030-23593-2_4

Chapter 2
Methods and Models for Information Data Analysis

2.1 Linear Random Processes

Many information signals like vibrations, acoustic emission signals, control signals, etc., can be represented as a response of some linear system on the white noise action [1, 2].

Now we try to make more exact the white noise models and make more detailed mathematical descriptions, definitions, and practical applications of the white noise processes.

The first detailed investigation of white noise processes was made by Ito in 1954. However, the beginning of the white noise theory can be traced back to the 1930s to the works of Kolmogorov and Khinchin [1], where the processes with independent increments closely related to white noise were considered.

In the most simple case, in technical applications, white noise can be defined as a generalized process with non-correlated values $\{\varsigma(t), t \in T\}$ such that

$$M[\varsigma(t)] = 0,\ M[\varsigma(t)\varsigma(t+\tau)] = \kappa_2\delta(\tau), \tag{2.1}$$

where $\delta(\tau)$ is a Dirac delta function. The stochastic process is generalized and stationary. The correlation function of the process is also a generalized one and is determined as

$$R(\tau) = \kappa_2\delta(\tau),\ \tau \in T. \tag{2.2}$$

where $\kappa_2 > 0$ is an intensity of the white noise. In the case of the nonstationary white noise, the parameter κ_2 will depend on time t.

White noise allows a spectral decomposition

$$\varsigma(t) = \frac{1}{2\pi}\int_{-\infty}^{\infty} e^{i\omega t} dz(\omega),$$

V. P. Babak et al., *Diagnostic Systems For Energy Equipments*, Studies in Systems, Decision and Control 281, https://doi.org/10.1007/978-3-030-44443-3_2

where $z(\omega)$ is a process with non-correlated increments.

The infinitesimal "elementary oscillations" $e^{i\omega t}dz(\omega)$ have equal infinitesimal mean power and are mutually non-correlated at any frequencies (ω) because of noncorrelated increments of the process $z(\omega)$. The mean square of their infinitesimal amplitude has a mathematical expectation

$$M\left|dz(\omega)\right|^2 = \kappa_2 d\omega, \quad \omega \in (-\infty, \infty).$$

The Gilbert stochastic process with representation (2.2) is called Loeve harmonized process. However, the white noise cannot be considered as the Loeve harmonized stochastic process, because it is not a Gilbert process: it does not have finite variance or finite mean power. Nevertheless, in a general sense, the white noise can be considered as a harmonized one. Then, its generalized harmonized correlation function can be derived from (2.2) and is determined as

$$R(\tau) = \frac{1}{2\pi}\int\limits_{-\infty}^{\infty} \kappa_2 e^{-i\omega\tau} d(\omega). \tag{2.3}$$

Thus, the white noise with continuous time has constant spectral power density at infinite frequency bound and equals to

$$S(\omega) = \kappa_2, \ \ \omega \in (-\infty, \infty),$$

i.e. the power of every infinitesimal harmonic component equals to $\kappa_2 d\omega$.

For the processes with continuous time the values of the white noise process in the strong sense and of the process with independent increments are connected by an equation

$$\eta(t) = \int\limits_0^t \varsigma(\tau)d\tau, \ t \in [0, \infty],$$

where $\{\varsigma(t), t \in [0, \infty)\}$ is a white noise in the strong sense with continuous time. The increments $\eta(\tau)$ at the non-crossing intervals are independent. If there exist a difference limit for $\Delta^s\eta(\tau) = \eta(\tau) - \eta(s)$, i.e.,

$$\varsigma_\tau = \lim_{|\tau - s| \to 0} \frac{\Delta^s\eta(\tau)}{\tau - s},$$

then the limit process ς_τ is the white noise process in the strong sense with continuous time. However, in a general sense, such a derivative does not exist and the white noise with continuous time can be considered as a generalized process.

In contrast to the white noise with continuous time, the processes with independent increments are physically possible.

Homogeneous characteristic function of the process with independent increments $\{\eta(t), t \in T\}$ is determined as

$$f_\eta(u,t) = Me^{iu\eta(t)} \tag{2.4}$$

The one-dimensional characteristic function $f(u; s, \tau)$ of increments $\Delta^s\eta(\tau)$ is determined as

$$f(u; s, \tau) = Me^{iu\Delta^s\eta(t)}.$$

A remarkable feature of the characteristic function (2.4) of the homogeneous processes with independent increments is that the corresponding distribution function belongs to the class of infinitely divisible ones, as described by Levy. A canonical representation of the characteristic function in Levy form is

$$\ln f_\eta(u,t) = |t|\left\{i\mu\varepsilon u - \frac{\delta^2}{2}u^2 + \int_{-\infty}^{\infty}\left[e^{iu\varepsilon x} - 1 - \frac{iu\varepsilon x}{1+x^2}\right]dL(x)\right\},\ t \in (-\infty, \infty), \tag{2.5}$$

where μ and $\delta > 0$ are some constants, $\varepsilon = signtL(x)$ is a Poisson jump spectrum in Levy form. It is noteworthy that $f_\eta(u,t) \equiv f(u; 0, t)$.

As was mentioned, the white noise with continuous time is a generalized stochastic process and it is not described by a distribution function, as such a function does not exist. The notions a Gaussian or a Poisson white noise indicate that the integral between boundary time interval has corresponding Normal or Poisson distribution, respectively.

White noise forms the basis of linear stochastic processes. Now we discuss a problem of statistical simulation of such signals. The problem, in general wording, consists in obtaining of a sequence of pseudo random values with the given probability characteristics. The proposed problem solving approach is based on the Linear Stochastic Process Theory [1] and it can be considered to be a development of innovation process method [3]. At first we state some knowledge of the Theory, and classify the simulating processes. Then we define concretely the simulation problem statement and discuss a method of solving.

A linear stochastic process (LSP) is a functional of the following form

$$\xi(t) = \int_{-\infty}^{\infty} \phi(\tau, t)d\eta(\tau),\ t \in (-\infty, \infty), \tag{2.6}$$

where $\phi(\tau, t) \in L_2(-\infty, \infty)$ with respect to τ for all t is a non-stochastic real Hilbert function; $\eta(\tau)$, $\eta(0) = 0$, $\tau \in (-\infty, \infty)$ is a Hilbert stochastically continuous random process with independent increments that is often called a generating process.

While solving many problems including statistical simulation it is convenient to consider the LSP as a response of a linear filter with the impulse transient function $\phi(\tau, t)$ on the action of the white noise $\eta'(\tau)$. It is understood that the $\eta'(\tau)$ is a generalized derivative of the corresponding process with independent increments.

As is shown in the paper [1], the LSP is an infinitely divisible process, that is, its finite-dimensional characteristic functions (CFs) are infinitely divisible ones. Therefore, the CF can be represented in one of the three canonical forms. Since the process $\xi(t)$ is the Hilbert one, we use the Kolmogorov representation form

$$f_{\xi}(u;t) = \exp\left\{\left[ium_{\xi}(t) + \int\limits_{-\infty}^{\infty} \{e^{ixu} - 1 - ixu\}\frac{dK_{\xi}(x,t)}{x^2}\right]\right\}, \tag{2.7}$$

where $m_{\xi}(t) = \mathbf{M}\xi(t)$; $K_{\xi}(x, t)$ is a jumps' spectrum in Kolmogorov form. We remark, the CF $f_{\eta}(u, t)$ of the process $\eta(\tau)$ can be represented analogously because it is infinitely divisible too [4]. The $f_{\eta}(u, \tau)$ is defined uniquely by pair $\left(m_{\eta}(\tau), K_{\eta}(x, \tau)\right)$.

Extending the results of the paper [3] to the non-stationary case we obtain expressions, which determine the process $\xi(t)$ characteristic's relation to the corresponding ones of the generating process $\eta(\tau)$:

$$m_{\xi}(t) = \int\limits_{-\infty}^{\infty} \phi(\tau, t)dm_{\eta}(\tau);\ K_{\xi}(x,t) = \int\limits_{-\infty}^{\infty}\int\limits_{-\infty}^{\infty} \phi^2(\tau,t)U(x - z\phi(\tau,t))d_{z\tau}K_{\eta}(z,\tau), \tag{2.8}$$

where $U(s)$, $s \in (-\infty, \infty)$ is the Heaviside function.

LSPs cover a wide class of stochastic processes. It is convenient to divide the class into subclasses and to introduce the corresponding terminology. As it is shown from (2.6) LSPs can be classified by the two main features: type of the generating process $\eta(\tau)$ and type of the function $\phi(\tau, t)$. Let us consider the matter in detail.

It is known [1] any stochastically continuous process with independent increments can be represented as a sum of two stochastically independent components which may be not present simultaneously: Gaussian and Poisson. We call the components as processes of Gaussian and Poison types. The first type contains the homogeneous (Wiener) and non-homogeneous Gaussian processes with independent increments. Simple Poisson processes, renewal processes and their linear combinations, generalized Poisson processes with independent increments belong to the second type. Being generated by each of the mentioned processes LSPs possess some typical properties. This fact is taken as a principle of the classification. For example, it can be shown that if the generating process is of the Gaussian type, the corresponding LSP is a Gaussian stochastic process. Poisson type of generating process leads to LSPs describing impulse currents. We call such processes as impulse LSPs.

The type of integral representation kernel $\phi(\tau, t)$ also influences typically the probability properties of the resulting process. As has been mentioned above, the physical meaning of some forming filter's impulse transient function can be attached to the $\phi(\tau, t)$. For example, the exponential kernel corresponds to the case of a low pass filtration, and the exponential cosine kernel corresponds to the band pass filtration case. By analogy with the forming filter's type we often call the corresponding LSPs as RC- and RLC-noises.

Thus, we here reveal briefly the main principle of LSPs' classification. The detailed discussion will be given in the paper.

Suppose, the signal being modelled is described by the LSP (2.6), and the latter is defined by pair $m_\xi(t)$, $K_\xi(x, t)$. Then the statistical simulation problem can be formulated in the following way.

Statement. We have to obtain a pseudo random sequence of values $\xi(k\Delta t)$, $k \in \mathbf{Z}$, $\Delta t \subset (-\infty, \infty)$, if the mean $m_\xi(t)$, jumps' spectrum $K_\xi(x, t)$ and function $\phi(\tau, t)$ are given.

Solution. We notice immediately, that the process of solving is reduced to the determination of generating process $\eta(\tau)C$ characteristics. We restrict ourselves by the case $m_\xi(t) = m_\xi = \text{const}$; $K_\xi(x, t) = K_\xi(x)$ that takes place, in particular, when $\phi(\tau, t) \equiv \phi(s)$, $s = t - \tau$ and the generating process is homogeneous, that is $m_\eta(\tau) = m_\eta \cdot \tau$; $K_\eta(x, \tau) = |\tau| K_\eta(x)$. Let us make a preliminary computation.

Rewrite the expressions (2.8) with reference to the assumption

$$m_\xi = m_\eta \int_{-\infty}^{\infty} \phi(s) ds; \; K_\xi(x) = \int_{-\infty}^{\infty} \int_{-\infty}^{\infty} \phi^2(s) U(x - z\phi(s)) ds \, dK_\eta(z). \tag{2.9}$$

Since $\int_{-\infty}^{\infty} \phi(s) ds \neq 0$, it follows from the first expression (2.9)

$$m_\eta = m_\xi / \int_{-\infty}^{\infty} \phi(s) ds. \tag{2.10}$$

The second of (2.9) can be represented in an operator form

$$K_x(x) = A K_h(z). \tag{2.11}$$

If the inverse operator A^{-1} exists, then

$$K_\eta(z) = \mathbf{A}^{-1} K_\xi(x). \tag{2.12}$$

The following reasoning concerns the function $K_\eta(z)$. It is known, if $K_\eta(z)$ has a jump at zero, that is

$$K_\eta(0+) - K_\eta(0-) = \sigma^2 \neq 0, \tag{2.13}$$

the Gaussian component of the process $\eta(\tau)$ is present. By force of the above supposition, the σ^2 is finite. Therefore, the jump at zero can be eliminated. Following Kolmogorov [1], we introduce a continuous at zero function

$$\bar{K}_\eta(z) = \begin{cases} K_\eta(z), \ z < 0; \\ K_\eta(z) - \sigma^2, \ z \geq 0. \end{cases} \tag{2.14}$$

If simultaneously $\sigma^2 \neq 0$ and $\bar{K}_\eta(z) \equiv 0$ then the generating process contains both components. Let us consider this general case.

From the above mentioned we can write

$$\eta(\tau) = w(\tau) + \pi_1(\tau), \ \tau \in (-\infty, \infty), \tag{2.15}$$

where $\{w(\tau), w(0) = 0\}$ is the Gaussian process with independent increments; $\{\pi_1(\tau), \pi_1(0) = 0\}$ is the generalized Poisson process with independent increments. Taking into account the stochastic independence of the processes $w(\tau)$ and $\pi_1(\tau)$, the characteristic function $f_\eta(u; \tau)$ is represented as a product

$$f_\eta(u; \tau) = f_w(u; \tau) \cdot f_\pi(u; \tau), \tag{2.16}$$

where

$$f_w(u; \tau) = \exp\left\{|\tau|\left[i\hat{u}m_w - \frac{\sigma^2 u^2}{2}\right]\right\} \tag{2.17}$$

is the CF of the Gaussian component with $\mathbf{M}w(\tau) = m_w\tau$; $\mathbf{D}w(\tau) = \sigma^2|\tau|$;

$$f_\pi(u; \tau) = \exp\left\{|\tau|\left[i\hat{u}m_\pi + \int_{-\infty}^{\infty}\left(e^{i\hat{u}z} - 1 - i\hat{u}z\right)\frac{d\bar{K}_\eta(z)}{z^2}\right]\right\}, \tag{2.18}$$

is the CF of the Poisson component with $\mathbf{M}\pi_1(\tau) = m_\pi\tau$; $\mathbf{D}\pi_1(\tau) = \bar{K}_\eta(\infty)|\tau|$; moreover $m_\eta = m_w + m_\pi$; $\hat{u} = u \cdot sign(\tau)$.

The simulation of the Gaussian component is not difficult if the m_w and σ^2 are known. As for one of the generalized Poisson process, it is necessary to have a jump's distribution function $F(y)$ and intensity λ of the jumps.

From distribution function's properties it follows that $\lambda = F_1(\infty)$, then

$$F(y) = F_1(y)/\lambda, \tag{2.19}$$

We can now calculate

$$m_x = \lambda \int_{-\infty}^{\infty} ydF(y) = \int_{-\infty}^{\infty} ydF_1(y),\ m_w = m_\eta - m_x. \tag{2.20}$$

Thus, the necessary characteristics for statistical simulation of the process $\xi(t)$ as a response of linear filter on the known action are found.

2.2 Linear AR and ARMA Processes

Now we consider a problem of statistical simulation of discrete time linear random processes (LRPs). One of the most interesting models of such class is a linear autoregressive process (AR-process) and linear autoregressive moving-average process (ARMA-process). We show some methods of the process's representation and formulate the simulation problem for this case.

2.2.1 Kernels of Linear AR and ARMA Processes

The linear stationary AR-process, by definition, can be represented as

$$\xi_t = -\sum_{i=1}^{p} a_i \xi_{t-i} + \varsigma_t,\ t \in Z, \tag{2.21}$$

where $\{a_i, a_i \neq 0, i = \overline{1,p}\}$, are autoregressive parameters; p is the autoregressive order; $\{b_j, b_j \neq 0, j = \overline{1,q}\}$, b_j—are moving-average; q is the order of moving-average; ς_t is the generating process. It is an infinitely divisible process with independent values.

Linear stationary ARMA processes can be also represented

$$\xi_t = \sum_{\tau=1}^{\infty} \phi_{AR}(\tau)\varsigma_{t-\tau}, \tag{2.22}$$

where $\phi_{AR}(\tau)$ is a kernel of the linear process [5]. It is assumed that

$$\phi_{AR}(0) = 1;\quad \sum_{\tau=0}^{\infty} |\phi_{AR}(\tau)|^2 < \infty, \tag{2.23}$$

The kernel $\phi_{AR}(\tau)$ is recursively connected with autoregressive parameters [5]:

$$\phi_{AR}(0) = 1,\ \text{if}\, p = 1,\ \phi_{AR}(s) = -a_p \phi_{AR}(s-1);\ s = 1, 2, \ldots, \tag{2.24}$$

If $p > 1$:

$$\phi_{AR}(s) = -\sum_{j=1}^{s} a_j \phi_{AR}(s-j) \text{ if } s = \overline{1, p-1},$$

$$\phi_{AR}(s) = -\sum_{j=1}^{p} a_j \phi_{AR}(s-j) \text{ if } s = p, p+1, \ldots$$

where $\phi_{AR}(\tau)$—is a kernel of linear stationary AR-process

$$\phi_{AR}(\tau) = \begin{cases} 0 \; k < 0, \\ 1 \; k = 0, \\ \sum\limits_{\tau=1}^{k} a_\tau \phi_{AR}(k-\tau) \; k = \overline{1, p-1}, \\ \sum\limits_{\tau=1}^{p} a_\tau \phi_{AR}(k-\tau) \; k = \overline{p, p+1}. \end{cases} \tag{2.25}$$

A linear stationary ARMA process can be defined as

$$\xi_t = -\sum_{i=1}^{p} a_i \xi_{t-i} + \varsigma_t + \sum_{j=1}^{q} b_j \varsigma_{t-j}, \tag{2.26}$$

where $\{a_i, a_i \neq 0, i = \overline{1, p}\}$, are autoregressive parameters; p is the autoregressive order; $\{b_j, b_j \neq 0, j = \overline{1, q}\}, b_j$—are moving-average; q is the moving average order; ς_t is the generating process. It is an infinitely divisible process with independent values.

Linear stationary ARMA processes can be also represented

$$\xi_t = \varsigma_t + \sum_{\tau=1}^{\infty} \phi_{ARMA}(\tau) \varsigma_{t-\tau}, \tag{2.27}$$

where $\phi_{ARMA}(\tau)$—is a kernel of linear stationary ARMA process

$$\phi_{ARMA}(k) = \begin{cases} 0 \, k < 0, \\ 1 \, k = 0, \\ b_k + \sum\limits_{\tau=1}^{k} a_\tau \phi_{ARMA}(k-\tau) \; k = \overline{1, p-1}, \\ b_k + \sum\limits_{\tau=1}^{p} a_\tau \phi_{ARMA}(k-\tau) \; k = \overline{p, p+1}. \end{cases} \tag{2.28}$$

It is shown in the papers [4, 6, 7] that the linear AR processes and Linear ARMA processes is an infinitely divisible process, that is, its finite-dimensional characteristic functions (CFs) are infinitely divisible ones. Therefore, the CF can be represented in one of the three canonical forms.

Linear autoregressive process with periodic structures $\{\xi_t, t \in Z\}$ defined over the set of integers $Z = \{\ldots, -1, 0, 1, \ldots\}$ are represented in paragraph 4.

2.2.2 Characteristic Functions of Linear AR and ARMA Processes

The process ξ_t is assumed to be strictly stationary, and adheres to the ergodic theorem [4, 8].

The process ξ_t has a Kolmogorov representation one-dimensional form logarithm of characteristic function (CF):

$$\ln f_\xi(u,t) = \ln f_\xi(u,1) = im_\xi u + \int_{-\infty}^{\infty} \left\{e^{iux} - 1 - iux\right\} \frac{dK_\xi(x)}{x^2}, \tag{2.29}$$

where parameter m_ξ and spectral functions of jumps $K_\xi(x)$ define unequivocally the characteristic function.

The logarithm of the one-dimensional characteristic function of the linear stationary autoregressive process can be written in the form

$$\ln f_\xi(u,t) = \ln f_\xi(u,1) = im_\varsigma u \sum_{\tau=-\infty}^{\infty} \phi(\tau) + \sum_{\tau=-\infty}^{\infty} \int_{-\infty}^{\infty} \left(e^{iux\phi(\tau)} - 1 - iux\phi(\tau)\right) \frac{dK_\varsigma(x)}{x^2}, \tag{2.30}$$

where parameters m_ς and $K_\varsigma(x)$ define the characteristic function of the generative process ς_t while $\phi(\tau)$ is the kernel of the linear random process ξ_t. The parameters m_ξ and m_ς, and Poisson spectra of jumps $K_\xi(x)$, $K_\varsigma(x)$ are interrelated as follows

$$m_\xi = m_\varsigma \sum_{\tau=0}^{\infty} \phi(\tau), \; K_\xi(x) = \int_{-\infty}^{\infty} R_\phi(x,y) dK_\varsigma(y), \tag{2.31}$$

where $R_\phi(x,y)$ is so-called transformation kernel, which is invariant with generative process ς_t and uniquely defined by the coefficients $\left\{a_j, a_j \neq 0, j = \overline{1,p}\right\}$ or $\left\{a_i, a_i \neq 0, i = \overline{1,p}\right\}$ and $\left\{b_j, b_j \neq 0, j = \overline{1,q}\right\}$.

Singularities of the $R_\phi(x,y)$ are discussed in the papers [4, 9]. Inverse kernel $R_\phi^{-1}(x,y)$ exist and the inverse integral transform exist also:

$$K_\varsigma(y) = \int_{-\infty}^{\infty} R_\phi^{-1}(x, y) dK_\xi(x). \quad (2.32)$$

Sometimes, application require finding statistical characteristic of the generating process ς_t in autoregressive parameters $\{a_j, a_j \neq 0, j = \overline{1, p}\}$ or $\{a_i, a_i \neq 0, i = \overline{1, p}\}$ and moving average parameters $\{b_j, b_j \neq 0, j = \overline{1, q}\}$ are known. Statistical characteristic of observed Linear AR process or linear ARMA process are known. Occasionally, such a problem is referred to as an inverse problem.

It is known [1] any stochastically continuous process with independent increments can be represented as a sum of two stochastically independent components which may be not present simultaneously: Gaussian and Poisson. We call the components as processes of Gaussian and Poison types. The first type contains the homogeneous (Wiener) and non-homogeneous Gaussian processes with independent increments. Simple Poisson processes, renewal processes and their linear combinations, generalized Poisson processes with independent increments belong to the second type. Being generated by each of the mentioned processes LRPs possess some typical properties. The statement could be generalized to the random process with discrete time. This fact is taken as a principle of the classification and simulation algorithms

2.3 Linear Random Processes with Periodic Structures

Numerous phenomena in radio engineering, mechanical, and biophysical systems as well as their respective random functions exhibit characteristics which are repeated in time and space. The examples of such processes are shot currents in electronic tubes, noise in a periodic pulse generator, magnetic noise in cyclic magnetization of ferromagnetics, and signals at the output of linear systems with periodically changing parameters such as parametric amplifiers with periodic pumping.

Mathematical models describing physical phenomena with pronounced periodic properties may take on the form of periodic, quasi-periodic, and periodically correlated random functions. A special case of periodic random processes is a class of stationary processes whose correlative theory was developed by Khinchin [1].

Further development of the theory of periodically correlated random processes as applied to description of modulated signals and processes occurring in radio engineering systems as well as for description of physical effects in acoustics and hydroacoustics is reflected in [2, 10–13].

To describe periodic processes in linear systems the deterministic approach is of common use since trigonometric functions are proper functions of any linear system.

The random periodic processes studied by Slutskiy [10] may be defined as follows.

According to Slutskiy, a real random process specified in some probabilistic space $\{\Omega, F, P\}$ is a random process $\xi(t), \quad t \in (-\infty, \infty)$ for which such $T > 0$ exists that the finite dimensional vectors $\Xi_1 = (\xi(t_1), \xi(t_2), \ldots, \xi(t_n))$ and $\Xi_2 = (\xi(t_1 + T), \xi(t_2 + T), \ldots, \xi(t_n + T))$ are stochastically equivalent, in a wide sense, for all numbers $n > 0$, where $t_1, t_2, \ldots$ is a set of separability of the process $\xi(t)$. Recall that two random vectors are stochastically equivalent in wide sense if for every integer $n > 0$ their partial distributions are coinciding, i.e. $P\{\omega : (\Xi_1) \in B\} = P\{\omega : (\Xi_2) \in B\},\ \omega \in \Omega,\ B \in F$.

The main purpose of the paper is to make the periodic nonstationary random processes more exact in an infinitely divisible class by detailed mathematical investigation of the problem connected with the description, definition, and practical application of the processes.

According to [1, 2], for the time-continuous processes the values of white noise in a the restricted sense $\{\varsigma(\tau),\ P\{\varsigma(0) = 0\} = 1\},\ \ \tau \in (-\infty, \infty)$ and those of the process with independent increments $\{\eta(t),\ t \in (-\infty, \infty)\}$ are related by the equation

$$\eta(t) = \int_0^t \varsigma(\tau) d\tau,\ t \in (-\infty, \infty). \tag{2.33}$$

while $\varsigma(\tau)$ is stochastically equivalent to $\varsigma(-\tau)$.

In our case, $\eta(t)$ is a non-homogeneous Hilbert process with independent increments. These exists a $T > 0$ for the process $\eta(t)$, which is the basis for the following properties' fulfillment:

$$\begin{aligned} d\kappa_1(\tau) &= d\kappa_1(\tau + T);\ d\kappa_2(\tau) \\ &= d\kappa_2(\tau + T);\ d_x d_t L(x, \tau) = d_x d_t L(x, \tau + T), \forall t \in (-\infty, \infty), \end{aligned} \tag{2.34}$$

where $\kappa_1(t)$ and $\kappa_2(t)$ are cumulative functions of the process $\eta(t)$ while $L(x, t)$ is its Poisson's spectrum of discontinuities in Levi's formula. The T number is called a period while $\varsigma(\tau)$, a periodic white noise.

The logarithm of a characteristic function of a real random process $\eta(t)$ in Levi's terms has the form

$$\ln f_\eta(u, t) = \left\{ i\mu(t)\varepsilon u - \frac{D^2(t)}{2} u^2 + \int_{-\infty}^{\infty} \left[e^{iu\varepsilon x} - 1 - \frac{iu\varepsilon x}{1 + x^2} \right] d_x L(x, t) \right\},\ t \in (-\infty, \infty) \tag{2.35}$$

where $D^2(t)$, $\mu(t)$ are some variables, $\varepsilon = sign\ \ t$ and $L(x, t)$ is the Poisson spectrum of discontinuities in Levi's representation.

Having used a nonhomogeneous random process with independent increments $\eta(t)$, we may generate a random process in the form [1]

$$\xi(t) = \int_{-\infty}^{\infty} \phi(\tau, t) d\eta(\tau),\ \ t \in (-\infty, \infty). \tag{2.36}$$

where $\phi(\tau, t) \in L_{2,\kappa}$ is a real nonrandom numerical function of period t such that $\int_{-\infty}^{\infty} \phi^2(\tau, t) d\kappa_2(\tau) < \infty$, at each fixed $t \in (-\infty, \infty)$. $\{\eta(\tau), \eta(0), \ t \in (-\infty, \infty)\}$ is a so-called generative process for which relations (2.36) are satisfied. Generally speaking, the process $\xi(t)$ is a nonstationary random one. Extending to the complex-value case may be made in the usual fashion.

On the strength of the assumptions made and with the account of (2.34) and (2.35), the logarithm of the characteristic function of the Levi type corresponding to linear random process (2.36) becomes [3]:

$$
\begin{aligned}
\ln f_{\xi}(u, t) &= iu \int_{-\infty}^{\infty} \phi(\tau, t) d\mu(\tau) + \int_{-\infty}^{\infty} \phi^2(\tau, t) dD(\tau) \\
&+ \int_{-\infty}^{\infty} \int_{-\infty}^{\infty} \left[\exp^{ixu\phi(\tau,t)} - 1 - \frac{iux\phi(\tau,t)}{1+x^2}\right] d_x d_\tau L(x, \tau), \\
d\mu(\tau) &= d\kappa_1(\tau) - \int_{-\infty}^{\infty} \frac{x^3}{1+x^2} d_x d_\tau L(x, \tau), \\
dD(\tau) &= d\kappa_2(\tau) - \int_{-\infty}^{\infty} x^2 d_x d_\tau L(x, \tau),
\end{aligned}
\tag{2.37}
$$

and, on the strength of (2.34): $d\mu(\tau) = d\mu(\tau + T)$; $dD(\tau) = dD(\tau + T)$.

Since conditions (2.34) and (2.35) are met,

$$
\begin{aligned}
\int_{-\infty}^{\infty} \phi(\tau, t) d\mu(\tau) = \int_{-\infty}^{\infty} \phi(\tau, t+T) d\mu(\tau), \ \int_{-\infty}^{\infty} \phi^2(\tau, t) dD(\tau) + \int_{-\infty}^{\infty} \phi^2(\tau, t+T) dD(\tau), \\
\int_{-\infty}^{\infty} \int_{-\infty}^{\infty} \left[\exp^{ixu\phi(\tau,t)} - 1 - \frac{iux\phi(\tau,t)}{1+x^2}\right] d_x d_\tau L(x, \tau) = \\
\int_{-\infty}^{\infty} \int_{-\infty}^{\infty} \left[\exp^{ixu\phi(\tau,t)} - 1 - \frac{iux\phi(\tau,t+T)}{1+x^2}\right] d_x d_\tau L(x, \tau + T).
\end{aligned}
$$

Thus, an identity is valid such that

$$
f_{\xi}(u, t) = f_{\xi}(u, t + T). \tag{2.38}
$$

That the process $\xi(t)$ is of Hilbert's type becomes obvious from the squared integration of $\phi(\tau, t)$ and from Hilbert's nature of $\eta(t)$.

It is sometimes expedient to use for application purposes somewhat simpler models which may be regarded as a special case of process (2.36). Such models include:

(a) linear random processes with an invariant kernel $\phi(\tau, t) \equiv \phi(t - \tau)$ and a nonhomogeneous generating process $\eta(t)$, i.e. the processes which may be set as

$$
\xi_1(t) = \int_{-\infty}^{\infty} \phi_1(t - \tau) d\eta(\tau), t \in (-\infty, \infty);
$$

(b) stationary linear random processes determined as

$$\xi_2(t) = \int_{-\infty}^{\infty} \phi_2(t-\tau)d\eta_1(\tau), t \in (-\infty, \infty),$$

where $\eta_1(t)$ is a uniform Hilbert's random process with independent values. Such random process $\xi_2(t)$ periodic in Slutskiy's terms for each $T \in (-\infty, \infty)$;

(c) processes with periodic kernel and generating homogeneous Hilbert's random process $\eta_1(t)$. Such a process may be assigned as

$$\xi_3(t) = \int_{-\infty}^{\infty} \phi_3(t-\tau)d\eta_1(\tau),\ t \in (-\infty, \infty).$$

The kernel $\phi(\tau, t)$ for such processes has a properly that $\phi(\tau, t) \equiv \phi(\tau, t+T)$.

Note that periodic, random processes in Slutskiy's terms, are also periodically correlated random processes (PCRP). As a rule, PCRP were studied within the L2 theory limits, and their correlative and spectral structure is known in sufficient detail. The main disadvantage of PCRP is that for such processes the general expression of their characteristic function is difficult to write down. For linear random processes the canonical form of characteristic function is simpler since such processes arc set constructively on the random processes basis having infinitely divisible distribution patterns. However, for nonstationary processes with similar distribution patterns the spectral structure is not seen directly. Thus, to describe nonstationary random processes having periodic structure it is expedient to generate models which could combine the benefits of PCRP and the random processes having the infinitely divisible distribution patterns. Among those arc linear stochastically periodic processes (2.36) whose properties were discussed before.

Such models may find wide use, for instance, for solving modulation problems, and for describing the behavior of radio engineering devices subject to additive as well as multiplicative interference. Let us consider this issue more thoroughly,

Here the so-called "multiplicative" model should be mentioned which has been widely practiced in the modulation problems, measurement theory, error estimating, etc.

Let $\xi(t)$ be a random process describing the behavior of a diagnostic system or section while $f(t)$ is an arbitrary deterministic function taking no zero values over $t \in (-\infty, \infty)$. Then $\xi(t) = \xi_1(t) * f(t)$, where $\xi_1(t) = \xi_1(t)/f(t)$. Hence, the model $\xi(t) = \xi_1(t) * f(t)$ may be used for describing the behavior of the information signal if $\xi_1(t)$ is regarded as a multiplicative interference. By analogy, we may write: $\xi(t) = \xi_2(t) + f(t)$ where $f(t)$ is an arbitrary function of t, and $\xi_2(t) = \xi(t) - f(t)$. Thus, the abstract model both of additive and multiplicative type is of no use since it does not separate anything, i.e. from the mathematical viewpoint, any random process is an "additive" and "multiplicative" model at the same time. However, there are additive and multiplicative interferences. Having applied the theory of linear random processes, we may describe the action both of additive and multiplicative interference on radio engineering devices and sections.

Using a characteristic function of a linear random process permits to carry out a complete analysis of output signals of linear systems from the standpoint of nonstationary periodic impacts, i.e., to calculate moments and distribution functions with the account of the properties of a generating process $\eta(t)$.

The application of such a model permits to generate pseudorandom sequences with desired probabilistic characteristics in the class of infinitely.

2.4 Linear Autoregressive Processes with Periodic Structures

Stationary autoregressive processes have found wide application in solving different diagnostic problems [4, 6, 7]. However, many processes in radiophysics, radiolocation, telemetry, hydroacoustics, meteorology, astronomy, biomedical systems and, consequently, the random functions describing such processes possess the characteristics repeating in time or space. Such processes can be used to describe, for example, signals at the mixer output, where a periodic oscillation and stationary noise are applied to the mixer inputs; the signals of parametric amplifiers with repetitive pumping, signals of nonlinear self-oscillatory systems, output signals of linear systems with cyclically varying parameters, and magnetic noises during the cyclic ferromagnetic magnetization switching. By the period of process variation is usually meant a time or space interval of complete repetition of variations of process characteristics, though the proper values (realizations) of such random process may not have the same properties. In this case for mathematical simulation it is expedient to apply the random processes with periodic structures.

Now we deal with the consideration of singularities and specific properties of linear autoregressive processes with periodic structures. Such processes are the generalization of linear stationary autoregressive processes [6]. The peculiarity of the given processes is the possibility of their use for the description of non-Gaussian periodic random signals.

Real random process $\{\xi_t, t \in \mathrm{Z}\}$ defined over the set of integers $\mathrm{Z} = \{\ldots, -1, 0, 1, \ldots\}$ is called the autoregressive process with periodically varying autoregressive parameters. It can be written as follows:

$$\xi_t + a_1(t-1)\xi_{t-1} + \cdots + a_p(t-p)\xi_{t-p} = \varsigma_t, \quad (2.39)$$

where $a_p(1)$ are the autoregressive parameters alternating in time with the same period i.e. $a_1(t) = a_1(t+T), \ldots, a_p(t) = a_p(t+T); p > 0, p \in \mathrm{Z}, \forall t \in \mathrm{Z}$ is the order of autoregressive, ς_t is the random process with discrete time and independent values having an infinitely divisible distribution law.

Autoregressive process has also the so-called state space representation [14]:

$$\begin{aligned} X_{t+1} &= A_p(t)X_t + B\varsigma_t, \\ \xi_t &= CX_t + \varsigma_t, \end{aligned} \tag{2.40}$$

where X_t, is the state vector, $X_t = \left[\xi_t^p, \ldots, \xi_t^1\right]$,

$$\xi_t^{p-j} = \sum_{i=j+1}^{p} -a_i(t)\xi_{t-i+j},\ j = 0, 1, \ldots, p-1,$$

$$A_p(t) = \begin{bmatrix} -a_1(t) & -a_2(t) & \cdots & -a_{p-1}(t) & -a_p(t) \\ 1 & 0 & \cdots & 0 & 0 \\ \vdots & \ddots & \cdots & & \vdots \\ 0 & 0 & \cdots & 0 & 0 \\ 0 & 0 & \cdots & 1 & 0 \end{bmatrix},\ C = B' = [1\ldots 0\,0],$$

′—is the transposition symbol.

The Hilbert autoregressive process with periodically changing autoregressive parameters can be specified in the form:

$$\xi_t = \sum_{\tau=0}^{\infty} \phi(\tau, t)\varsigma_{t-\tau}, \tag{2.41}$$

where $\phi(\tau, t)$, $|\phi(\tau, t)| < K$ is the real or complex-valued function that is uniformly limited in terms of both arguments ($K < \infty$).

Therefore, the autoregressive process specified by Eq. (2.39) can be called linear discrete-time random process. Kernel $\phi(\tau, t)$ of linear random process (2.41) is related to autoregressive parameters $\left\{a_j(t), j = \overline{1,p}\right\}$ at the fixed value of t by the relationships presented in paper [12]. The kernel of linear random process (2.41) can be also determined by using the system of matrix Eq. (2.40):

$$\phi(\tau, t) = CA_p^{\tau-1}(t)B. \tag{2.42}$$

For the fixed value of t, $A_p^0 = I$, where I is the identity matrix, while $A_p^\tau = A_p{*}A_p^{\tau-1}$ [15].

From expression (2.40) it follows that the following relationships should be fulfilled for the autoregressive process with periodic coefficients:

$$A_p(t) = A_p(t+T),\ \phi(\tau, t) = \phi(\tau, t+T),\ T > 0. \tag{2.43}$$

Hence it follows that the following equalities are fulfilled:

$$CA_p(t)B = CA_p(t+T)B,\ \phi(\tau, t) = \phi(\tau, t+T),\ T > 0.$$

Thus, the Hilbert autoregressive process with cyclically time varying autoregressive parameters having the same period $T > 0$ generated by a random process with independent values and completely divisible distribution law represents a linear random process with discrete time and periodic kernel (in terms of t).

The logarithm of characteristic function of the linear autoregressive process with periodic kernel has the following form:

$$\begin{array}{c} \ln f_{\xi}(u,t) = i\kappa_{\varsigma 1}u\sum\limits_{\tau=0}^{\infty}\phi(\tau,t) - 0.5u^2\kappa_{\varsigma 2}\sum\limits_{\tau=0}^{\infty}\phi^2(\tau,t) + \\ + \sum\limits_{\tau=0}^{\infty}\int\limits_{-\infty}^{\infty}\left\{\exp[ixu\phi(\tau,t)] - 1 - \frac{iux\phi(\tau,t)}{1+x^2}\right\}dL(x) \end{array}, \quad (2.44)$$

where $\kappa_{\varsigma 1}, \kappa_{\varsigma 2}$ are the first and second semi-invariants of the generating process ς_τ; respectively, $L(x)$ is the Poisson spectrum of jumps in the Levi formula of the generating random process ς_τ.

Let us consider a linear autoregressive process having the periodic structure of the generating process. Such process satisfies the following difference equation:

$$\xi_t + a_1\xi_{t-1} + \cdots + a_{t-p} = \varsigma_t,\ t \in \mathrm{Z}, \quad (2.45)$$

where $\left\{a_j, j = \overline{1,p}\right\}$ are the autoregressive parameters, p is the order of autoregressive, ς_t is the generating process having the properties presented below.

Let us assume that $\eta_t = \varsigma_t - \varsigma_{t-1},\ t \in \mathrm{Z}$ is the first difference of generating process ς_t. Let us also assume the existence of such $T > 0$ that for all τ and t the following relationships are fulfilled:

$$\kappa_1(\tau) = \kappa_1(\tau+T),\ \kappa_2(\tau) = \kappa_2(\tau+T),\ d_xL(x,\tau) = d_xL(x,\tau+T), \quad (2.46)$$

where $\kappa_1(\tau)$ and $\kappa_2(\tau)$ are the first cumulant functions of process η_t; $L(x,\tau)$ is the Poisson spectrum of jumps in the Levi formula for process ς_t.

The logarithm of characteristic function of the linear random process with periodic generating process has the following form:

$$\begin{array}{c} \ln f_{\xi}(u,t) = iu\sum\limits_{\tau=0}^{\infty}\phi(t-\tau)\mu(\tau) - 0.5u^2\sum\limits_{\tau=0}^{\infty}\phi^2(t-\tau)D(\tau) + \\ + \sum\limits_{\tau=0}^{\infty}\int\limits_{-\infty}^{\infty}\left\{\exp[ixu\phi(t-\tau)] - 1 - \frac{iux\phi(t-\tau)}{1+x^2}\right\}d_xL(x,\tau) \end{array}, \quad (2.47)$$

where $\mu(\tau) = \kappa_1(\tau) - \int_{-\infty}^{\infty}\frac{x^3}{1+x^2}d_xL(x,\tau)$, $D(\tau) = \kappa_2(\tau) - \int_{-\infty}^{\infty}x^2 d_xL(x,\tau)$.

It can be seen that the following conditions are fulfilled [16]:

$$\mu(\tau) = \mu(\tau+T),\ D(\tau) = D(\tau+T).$$

Hence, the following identity is true:

$$f_{\xi}(u,t) = f_{\xi}(u,t+T). \quad (2.48)$$

Thus the linear autoregressive process with periodic generating process is a periodic random process in strict sense.

The above considered models of random processes can be applied for the simulation of information signals of expert systems, various cyclically changing radiophysical processes, periodic signals in biomedical investigations, etc.

The autoregressive processes with cyclically time-varying autoregressive parameters having the same period and with periodic generating process can be also referred to periodic random processes in strict sense.

Let us consider such autoregressive processes in a greater detail.

Suppose we have the following autoregressive process

$$\xi_t + a_1(t-1)\xi_{t-1} + \cdots + a_p(t-p)\xi_{t-p} = \varsigma_t, t \in \mathrm{Z}. \quad (2.49)$$

where $a_1(t) \ldots a_p(t)$ are the autoregressive parameters cyclically varying with time and having the same period $T_1 > 0$, i.e.,

$$\begin{aligned} a_1(t) &= a_1(t+T_1) \\ &\vdots \\ a_p(t) &= a_p(t+T_1), \end{aligned}$$

where $p > 0$, $p \in \mathrm{Z}$ is the order of autoregressive.

Let $\eta_t = \varsigma_t - \varsigma_{t-1}$, $t \in \mathrm{Z}$ be the first difference of generating process ς_t. Such $T_2 > 0$ that for all τ and t the following relationships are fulfilled:

$$\kappa_1(\tau) = \kappa_1(\tau + T_2),\ \kappa_2(\tau) = \kappa_2(\tau + T_2),\ d_x L(x,\tau) = d_x L(x, \tau + T_2). \quad (2.50)$$

where $\kappa_1(\tau)$ and $\kappa_2(\tau)$ are the first cumulant functions of process η_t; $L(x,\tau)$ is the Poisson spectrum of jumps in the Levi formula for process ς_t. Let us assume the existence of such real number a $\alpha \in (-\infty, \infty)$ that $T_2 = \alpha T_1$. Let us designate $T_1 = T$, then $T_2 = \alpha T$. Therefore, the kernel of linear random autoregressive process (2.49) satisfies relationship $\phi(\tau, t) = \phi(\tau + \alpha T, t + T)$ at all values of τ, t. Parameter α represents the ratio $\alpha = tg\varphi$ of the period of the first difference of generating process η_t, to the period of linear autoregressive ξ_t.

Process parameter α is determined by relationship $\alpha = tg\varphi$, where $tg\varphi$ is the slope ratio of plane $\tau \times t$, along which kernel $\phi(\tau, t)$ is a periodic function [17]. Angle φ is called the angle of kernel periodicity of process ξ_t.

Applying the linear autoregressive processes with periodic structures as mathematical models of information signals having the cyclic character, it is possible to construct effective algorithms of analysis and classification of such signals using a priori information about these processes. As compared with other models, having periodic structure and presented, for example, in papers [12] that can be applied for stochastic Gaussian information signals, the linear autoregressive processes with periodic structures make it possible to describe a wide class of stochastic nonstationary non-Gaussian information signals having completely divisible distributions and periodic structure. The use of such models can be effective in simulating the signals of passive radiolocation and hydrolocation systems, deep space communication, micro-satellite control, and diagnostics making it possible to detect defects at early stages of their appearance.

2.5 Inverse Problem of AR Processes

Autoregressive processes are widely practiced when constructing mathematical models of information signals of different types and during their analysis and synthesis. To classify such processes their energy characteristics are often used, but considering problems of classification of stochastic information signals in case of non-Gaussian distribution, the information, which energy spectra (within the framework of the first two moments) possess, is often not enough for reliable recognition and classification of such signals. Then it is expedient to use information on higher moments (integral characteristics) or statistical characteristics of such signals distributing.

Sometimes, applications require finding statistical characteristics of the generating process ς_t if autoregressive parameters are known $\{a_j, j = \overline{1, p}\}$, where p is the autoregressive order and statistical characteristics of the process observed ξ_t. Occasionally, such a problem is referred to as an inverse problem. For autoregressive processes such a problem was originally considered by Cox, where statistical characteristics of the generating process ς_t, were determined for die case of autoregressive processes of the first order $\xi_t + a_1\xi_{t-1} = \varsigma_t, t \in Z$ that further on will be designated as AR (1), while $Z = \{\ldots, -1, 0, 1, \ldots\}$ is the set of integers. It was assumed that it has exponential distribution. A similar problem for AR case (1) of the process that has gamma and binomial distribution was considered by Lawrance and McKenzie. The investigation dealing with determinations of statistical characteristics of generating processes, namely, moments generating function, consider only AR processes (1). In addition, generating function of the generating process ς_t moments was determined for the case of distributions that belong to infinitely devisable distributions class. This class also encompasses exponential (Laplace) distribution, gamma distribution, negative binomial distribution, and Gaussian distribution.

The present part offers, when finding statistical characteristics of generating process, to use the apparatus of characteristic functions. The paper also considers the problem of finding characteristic function of generating process, if the autoregressive linear process is observed. The method, which is going to be used for solving this problem, could be considered as the development of generating process method proposed by Kolmogorov [1] and also developed in [1–3].

Let us consider some properties of autoregressive linear processes in more detail.

Autoregressive linear processes (2.21) could be set in the following way:

$$\xi_t + \sum_{j=1}^{p} a_j \xi_{t-j} = \varsigma_t,\ t \in Z, \tag{2.51}$$

where $\{a_j,\ a_j \neq 0,\ j = \overline{1,p}\}$ are the autoregressive parameters; p is the autoregressive order; $\{\zeta_t, t \in \mathrm{Z}\}$ is the stationary random process with discrete time and independent values that has the infinitely divisible distribution law $P\{\varsigma_0 = 0\} = 1$. This process is often referred to as the generating process.

It is proposed that the process ς_t is stationary in the narrow sense and ergodic theorems are fulfilled [4], i.e. it is assumed that

$$M\,|\xi_t|<\infty,\ \frac{1}{m^2}\sum_{t=0}^{m}\sum_{n=0}^{m} r(t,n) \to 0 \,\text{if}\, m \to \infty, \tag{2.52}$$

where $r(t,n) = \mathbf{M}[(\xi_t - \mathbf{M}\xi_t)(\xi_n - \mathbf{M}\xi_n)]$ is the correlation function of process ξ_t.

It is also assumed that solutions of the characteristic equation

$$\Psi(z) = a_p + a_{p-1}z + \cdots + a_1 z^p,$$

on the complex plane lie within the unit disk [18]. Then, difference Eq. (2.51) has the only stationary solution

$$\xi_t = \sum_{\tau=0}^{\infty} \phi(\tau)\zeta_{t-\tau},$$

where $\{\phi(\tau),\ \tau \in \mathrm{Z}\}$ is some numerical sequence, which is referred to as a pulse-transition function or kernel of the random process ξ_t. It is assumed that the following relationship is fulfilled

$$\sum_{\tau=0}^{\infty} \phi^2(\tau) < \infty.$$

Consequently, the autoregressive linear process ξ_t of order p could be considered as the process of sliding average infinite order. It was shown in 2.1 that a kernel $\varphi(\tau)$ is related to the parameters of autoregressive $\{a_j,\ j = \overline{1,p}\}$ with recurrent forms.

To find the characteristic function of generating process ξ_t of the autoregressive linear process we propose a method that uses properties of Poisson spectra of jumps of the characteristic functions of infinitely-devisable laws distribution [1]. This approach is a development of Bruno de Finetti and Kolmogorov method. The following prerequisites form the basis of this method.

The logarithm of one-dimensional characteristic function for linear stationary process of autoregressive could be determined in Kolmogorov canonical form

$$\ln f_{\xi}(u,t) = \ln f_{\xi}(u,1) = im_{\xi}u + \int_{-\infty}^{\infty} \{e^{iux} - 1 - iux\} \frac{dK_{\xi}(x)}{x^2}, \tag{2.53}$$

where parameter m_{ξ} and spectral function of jumps $K_{\xi}(x)$ unambiguously determine the characteristic function.

The logarithm of autoregressive linear process characteristic function could be written down also in the following form [1]

$$\ln f_{\xi}(u,t) = \ln f_{\xi}(u,1) = im_{\zeta}u \sum_{\tau=-\infty}^{\infty} \phi(\tau) + \sum_{\tau=-\infty}^{\infty} \int_{-\infty}^{\infty} \left(e^{iux\phi(\tau)} - 1 - iux\phi(\tau)\right) \frac{dK_{\zeta}(x)}{x^2}, \tag{2.54}$$

where parameters m_{ς}, $K_{\zeta}(x)$ determine characteristic function of the generating process ς_t, while $\phi(\tau)$ is the kernel of linear random process in relationships (2.55) and (2.56) $K_{\xi}(x)$, $K_{\zeta}(x)$ are non-decreasing limited functions such that $K_{\varsigma}(-\infty) = 0$.

Thus, if the characteristic function of autoregressive stationary linear process ς_t is observed in canonical form (2.52) and kernel $\phi(\tau)$ is known, then using relationship (2.55) and considering results presented in papers [1, 6] one could determine the parameters m_{ς}, $K_{\zeta}(x)$, and construct the characteristic function of generating process ς_t in Kolmogorov canonical form.

Parameters m_{ξ}, m_{ς} and $K_{\xi}(x)$, $K_{\zeta}(x)$ are related by the relationships

$$m_{\xi} = m_{\zeta} \sum_{\tau=0}^{\infty} \varphi(\tau),$$

or

$$m_{\xi} = m_{\varsigma} \left\{ 1 + \sum_{\tau=1}^{\infty} \left[-a_p \phi(\tau - 1) \right] \right\}, \text{ if } p = 1.$$

If $p > 1$

$$m_{\xi} = \begin{cases} m_{\varsigma} \left\{ 1 + \sum\limits_{\tau=1}^{p} \left[-\sum\limits_{j=1}^{\tau} a_j \phi(\tau - j) \right] \right\} & \text{given } \tau = \overline{1, p-1}, \\ m_{\varsigma} \left\{ 1 + \sum\limits_{\tau=p}^{\infty} \left[-\sum\limits_{j=1}^{p} a_j \phi(\tau - j) \right] \right\} & \text{given } \tau = p, p+1, \ldots \end{cases}.$$

Poisson spectra of jumps $K_{\xi}(x)$ and $K_{\varsigma}(y)$ are related [1] (assuming that processes ξ_t and ς_t are stationary) as

$$K_{\xi}(x) = \int_{-\infty}^{\infty} R_{\phi}(x, y) dK_{\varsigma}(y), \tag{2.55}$$

where $R_{\phi}(x, y)$ is the transformation kernel. It is assumed that the relationship $\sum_{t=-\infty}^{\infty} \iint \left| R_{\phi}(x, y, t) \right|^2 dxdy < \infty$ is fulfilled and that inverse kernel $R_{\phi}^{*}(x, y)$ and inverse integral transformation exist

$$K_{\varsigma}(y) = \int_{-\infty}^{\infty} R_{\phi}^{*}(x, y) dK_{\xi}(x). \tag{2.56}$$

Relationships (2.55) and (2.56) are true, if Poisson spectra of jumps of the processes .. and ς_t do not have jumps in zero (i.e., if the processes ξ_t and ς_t belong to the class of processes with infinitely divisible laws of distribution without Gaussian component. In other cases one should use the relationships

$$K_{\xi}(x) - K_{\xi}(0) = \int_{-\infty}^{\infty} R_{\phi}(x, y) dK_{\varsigma}(y),\ K_{\varsigma}(x) - K_{\varsigma}(0) = \int_{-\infty}^{\infty} R_{\phi}^{*}(x, y) dK_{\xi}(y), \tag{2.57}$$

where $K_{\xi}(0)$ and $K_{\varsigma}(0)$ are jumps in zero.

Having used the results presented in [1, 4], transformation kernel for the stationary linear autoregressive processes could be found as

$$R_{\phi}(x, y) = \sum_{\tau=-\infty}^{\infty} \phi^2(\tau) U\left[x - y\phi(\tau)\right],\ x, y \in (-\infty, \infty) \tag{2.58}$$

where $\phi(\tau)$ is the kernel of autoregressive linear random process; $U[.]$ is the Heavyside function. Transformation kernel properties $R_{\phi}(x, y)$ are related with the properties $\phi(\tau)$ and, thus, with the properties of autoregressive parameters $a_1, \ldots, a_p$. Relationship (2.57) could be written down in the following form:

If $p = 1$,

$$R_{\phi}(x, y) = \left\{1 + \sum_{\tau=1}^{\infty} \left[a_p \phi(\tau - 1)\right]\right\}^2 U\left\{x + y\left[1 + \sum_{\tau=1}^{\infty} \rangle a_p \phi(\tau - 1) \langle\right]\right\},\ \phi(0) = 1,$$

If $p > 1$

$$R_\phi(x,y)=\begin{cases}\sum\limits_{\tau=1}^{p-1}\left\{\left[\sum\limits_{j=1}^{\tau}a_j\phi(\tau-1)\right]^2U\left[x+y\sum\limits_{j=1}^{\tau}a_j\phi(\tau-1)\right]\right\} & \text{for } \tau=\overline{1,p-1},\\ \sum\limits_{\tau=p}^{\infty}\left\{\left[\sum\limits_{j=1}^{p}a_j\phi(\tau-1)\right]^2U\left[x+y\sum\limits_{j=1}^{p}a_j\phi(\tau-1)\right]\right\} & \text{for } \tau=p,p+1,\ldots\end{cases}$$

Thus, to find the algorithm of one-dimensional characteristic function of generating process ς_t of the linear stationary autoregressive process ξ_t if the logarithm of one-dimensional characteristic function $\ln f_\xi(u,1)$ and kernel $\phi(\tau)$ are known, one may use the following method:

(1) parameter m_ς of the logarithm of generating process characteristic function in Kolmogorov form for linear stationary autoregressive process is determined using the formula $m_\zeta=m_\xi/\left[\sum_{\tau=0}^{\infty}\phi(\tau)\right]$;
(2) having used (2.58) we find the transformation kernel $R_\phi(x,y)$;
(3) having determined $R_\phi^*(x,y)$ from relationship (2.56) we find the Poisson spectrum of jumps in Kolmogorov form $K_\varsigma(y)$ of generating process ς_t for the linear autoregressive stationary process ξ_t.

Thus, having determined parameters m_ς and $K_\varsigma(y)$ one may construct characteristic function of the generating process ς_t in Kolmogorov canonical form.

The method considered provides a possibility to construct characteristic function of the generating processes for linear stationary autoregressive processes, when the characteristic function in Kolmogorov canonical form is known. The method is not confined to finding the generating processes for linear autoregressive processes of higher orders.

Consider an example of the inverse problem solution.

Let the process ξ_t have a gamma-distribution characterizied by one-dimensional characteristic function

$$f_\xi(u,t)=(1-iu\theta)^{-b},\ \forall t\in Z;\ \theta>0;\ b>0. \tag{2.59}$$

The process is assumed to be strictly stationary, and adheres to the ergodic properties (2.52).

Consider the peculiar features of Poisson spectra of jumps determination for the autoregressive process of the second order, i.e., for the process $\xi_t+a_1\xi_{t-1}+a_2\xi_{t-2}=\varsigma_t$, having one-dimensional gamma-distribution (2.59). In this case Poisson spectrum of process jumps in Kolmogorov's formula could be determined as

$$K_\xi(x)=b\int\limits_0^x y\exp(-y/\theta)dy=\begin{cases}b\theta\left[\theta-(\theta+x)\exp(-x/\theta)\right], & x\ge 0\\ 0 & ,x<0\end{cases}$$

whence $dK_\xi(x)=bx\exp(-x/\theta)dx,\ x\ge 0$.

Consider the simplest (for our problem) case when autoregressive parameters $a_1,a_2<0$. It is assumed that the relationships $a_1^2+4a_2<0$, $|a_1+a_2|<1$,

$|a_1 - a_2| < 1, -1 < |a_2| < 1, |a_1| < 1 - a_2$ are satisfied. Then the function $\phi(\tau)$ decreases but remains positive while the transform Kernel $R_\phi(x, y)$, in conformity with (2.60), could be defined as

$$R_\phi(x, y) = \begin{cases} \sum_{\tau=-\infty}^{\infty} \phi^2(\tau) & 0 \le \phi(\tau)y < x, \\ 0 & \phi(\tau)y > x;\ x<0,\ y = 0. \end{cases}$$

Taking into account the convergence of $\sum_{\tau=-\infty}^{\infty} \varphi^2(\tau)$ the properties $R_\phi(x, y)$, $\phi(\tau)$, and the existence of transform inverse kernel $R_\phi^{-1}(x, y)$, we could write

$$R_\phi^{-1}(x, y) = \begin{cases} \left[\sum_{\tau=-\infty}^{\infty} \phi^2(\tau)\right]^{-1} & 0 \le \phi(\tau)y < x, \\ 0 \quad \phi(\tau)y > x; & x<0;\ y=0. \end{cases}$$

Consequently, based on (2.57), the Poisson spectrum of jumps of the generative process ς_t, could be defined as

$$K_\varsigma(y) = \begin{cases} b\left|\sum_{\tau=-\infty}^{\infty} \phi^2(\tau)\right|^{-1} \int_{0_+}^{y} x\exp(-x/\theta)dx, & y > 0, \\ 0, & y = 0, \end{cases}$$

whence

$$K_\varsigma(y) = \begin{cases} b\left|\sum_{\tau=-\infty}^{\infty} \phi^2(\tau)\right|^{-1} \{\theta - (\theta + y)\exp(-y/\theta)\}, & y > 0, \\ 0, & y = 0. \end{cases}$$

Then

$$dK_\varsigma(y) = \theta b\left|\sum_{\tau=-\infty}^{\infty} \varphi^2(\tau)\right|^{-1} \{y\exp(-y/\theta)\}dy,\ y > 0, \tag{2.60}$$

and the logarithm of the generative process characteristic function

$$\begin{gathered} \ln f_\varsigma(u; t) = |t| \ln f_\varsigma(u; 1) = \\ = i\theta b|t|\left|\sum_{\tau=-\infty}^{\infty} \phi(\tau)\right|^{-1} u + b\theta|t|\left[\sum_{\tau=-\infty}^{\infty} \phi^2(\tau)\right]^{-1} \int_0^{\infty} \{\exp(iyu) - 1 - iuy\}\frac{\exp(-y/\theta)}{y}dy, \\ t = Z. \end{gathered} \tag{2.61}$$

where $\theta > 0,\ b > 0,\ y > 0,\ |a_1| < 1 - a_2$.

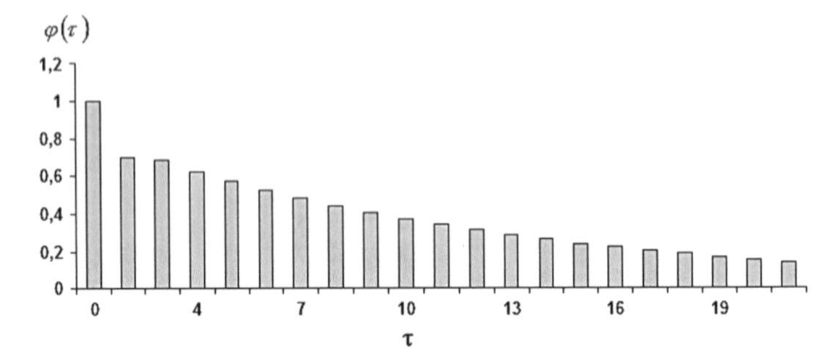

Fig. 2.1 The kernel of linear autoregressive process AR(2) that has parameters $a_1 = -0.7$ and $a_2 = -0.2$

After integration we could write

$$\ln f_{\varsigma}(u;t) = |t|\ln f_{\varsigma}(u;1) = \\ = i\theta b|t|\left|\sum_{\tau=-\infty}^{\infty}\phi(\tau)\right|^{-1}u + b\theta|t|\left[\sum_{\tau=-\infty}^{\infty}\phi^2(\tau)\right]^{-1}[iu(1-\theta) - \ln(1-iu\theta)]. \tag{2.62}$$

As an example let us consider an autoregressive process of the second order with coefficients $a_1 = -0.7$ and $a_2 = -0.2$,

$$\xi_t - 0,7\xi_{t-1} - 0,2\xi_{t-2} = \varsigma_t.$$

The process ξ_t has a gamma-distribution with parameters $\theta > 0,\ b > 0$, whose one-dimensional characteristic function could be represented by relationship (2.61).

In considered case the kernel of autoregressive linear random process $\phi(\tau)$ is a decreasing positive function. The kernel simulation results are represented in Fig. 2.1.

The kernel of autoregressive linear process (2.59) has parameters $a_1 = -0.7$ and $a_2 = -0.2$.

Having used mathematical simulation methods, we determined the parameter values of characteristic function (2.61)

$$\left[\sum_{\tau=0}^{n}\phi(\tau)\right]^{-1} = 0.1111 \quad \text{and} \quad \left[\sum_{\tau=0}^{n}\phi(\tau)^2\right]^{-1} = 0.29032$$

with the sampling volume $n = 1000$.

Then the logarithm of characteristic function for the generative process ς_t of autoregressive process (2.62) could be written as

$$\ln f_{\varsigma}(u;\, t) = |t|\ln f_{\varsigma}(u;\, 1) =$$
$$= 0.1111i|t|\theta bu + 0.29032\theta b|t| \int\limits_{0}^{\infty} \{\exp(iyu) - 1 - iuy\} \frac{\exp(-y/\theta)}{y} dy =$$
$$= b\theta|t|\{0.1111iu + 0.29032[iu(1-\theta) - \ln(1 - iu\theta)]\}.$$

The proposed method makes it possible to construct the characteristic function of generative processes for linear stationary autoregressive processes of the second order having gamma-distribution. The method is not limited only to autoregressive processes (2.61) and could be used for determination of generative processes for linear autoregressive processes of higher orders having infinitely divisible distributions. An example of the inverse problem solution for the AR (2) process that has negative binomial distribution was considered in [19].

2.6 Statistical Splines Application

Monitoring of current operating conditions of power engineering equipment with forecasting possible failures in such equipment is an issue of great importance. On the one hand, forecasting of possible equipment failures provides more reliable operation due to timely repair and replacement of the most worn parts. On the other hand, it helps to avoid significant financial expenses for complete replacement of equipment.

The most challenging way to solve this problem is implementation of special expert systems for monitoring of operating conditions and forecasting failures. Such systems on the basis of regular analysis of the actual equipment status allow prediction and avoiding of the failures appearance.

The usage of the methods of statistical diagnostics based on mathematical models of linear stochastic processes enables estimation of the actual state of the operating electrical equipment with high level of reliability.

Assuming the timely service, forecasting equipment failures using statistical spline-functions ensure a high level of its reliability and economical effectiveness in use.

The main idea is that some quantitative characteristics of physical processes running in different components of tested equipment change when failures or defects appear. This allows locating and identification of failures. Regular observation of gradual changes of such parameters values versus time may provide researcher with important information regarding the tendency of failure evolution and allow forecasting the possible time of the failure appearance.

2.6.1 Forecasting the Time of Failure Using Statistical Spline-Functions

In the scope of this paper let's consider only so-called degradation failures that are caused by the change of diagnostic object parameters due to materials aging, micro-defects accumulation etc. We will not consider stochastic failures, caused by unpredictable factors.

For the technical diagnostics of electrical equipment part, we need to choose some numerical diagnostic parameter (or a set of diagnostic parameters) and an appropriate criteria that provide a way to make a decision about technical status of given constructive part. These questions had been discussed in a number of scientific works, for example [20, 21]. Additionally, for the forecasting of the possible failure time of selected constructive part a statistics of diagnostic parameter values during some time range is necessary. We will call this time range as "observation interval".

Let's assume that studied electrical equipment operates in unchanging conditions during the whole observation interval $[a,\ b]$. In other words, no repair work is performed on studied equipment constructive part and routine maintenance does not change these conditions significantly.

In such case we can assume that the diagnostic parameter y vary in time t according to some functional dependence $y = f(t,\ A)$, where A is a determinate vector of unknown real parameters, that linearly appear in $f(t,\ A)$.

As a result of the series of experiments, the sequence of values $y_i, i = \overline{1,\ N}$ of the function y is obtained. Every y_i corresponds to the value of the argument $t_i \in [a,\ b]$, $i = \overline{1,\ N}$.

We can assume, that values y_i are distributed by Gaussian law with identical dispersions ($\mathbf{D}y_i = \sigma^2, i = \overline{1,\ N}$, where $\mathbf{D}$ is dispersion operator) and are uncorrelated. This comes from the fact that measurements of values y_i are obtained with the same measuring instrument with the same accuracy in the independent time moments.

Desired functional dependence could be represented in the following form:

$$\mathbf{M}y_i = \sum_{k=0}^{r} x_{ik} a_k = f(t_i,\ A),\ i = \overline{1,\ N}, \tag{2.63}$$

where $\mathbf{M}$ is distribution expectation operator; $A = (a_0, a_1, \ldots, a_r)$ are unknown parameters;$X = (x_{ik}),\ i = \overline{1,\ N},\ k = \overline{0,\ r}$—so-called planning matrix that consists of defined elements functionally dependent on t_i (not excluding the non-linear dependence).

Assuming that $Y = (y_i),\ i = \overline{1,\ N}$ and A are column matrixes, the Eq. (2.63) could be noted in the matrix form: $\mathbf{M}Y = XA$. Thus, the problem of dependence $y = f(t,\ A)$ reconstruction is reduced to the determination of statistical estimations of unknown parameters A from the results of observations $y_i,\ i = \overline{1,\ N}$. At that the elements of planning matrix could be arbitrary chosen. This allows the search of the optimal solution in some sense.

Note that elements of planning matrix define the class of functions that could be reconstructed. Let's discuss the solution of the problem in spline-functions class [20, 21]. Statistical estimations of unknown parameters will be constructed by the method of least squares.

Spline is a function, composed from pieces of different functions according to the defined scheme. Polynomial spline is composed from segments of different polynomials in a way that the resulting function is smooth enough. For the interpolation of some function with polynomial splines, a grid is defined on $[a,\ b]$ segment of t axis:

$$\Delta_r = \left\{t_j\right\}_{j=0}^{r},\ a = t_0 < t_1 < \cdots < t_r = b. \tag{2.64}$$

A function $S_m(t) = S_{m,k}(t,\ \Delta_r)$ is called polynomial spline of power m and defect k $(1 \le k \le m)$ with nodes (2.66) if

(a) $S_m(t) \in \mathbf{P}_m$ for $t \in \left[t_j,\ t_{j+1}\right], j = \overline{0,\ r-1}$,
(b) $S_m(t) \in \mathbf{C}^{m-k}[a,\ b]$,

where $\mathbf{P}_m$ is a set of real polynomials with power not exceeding m; $\mathbf{C}^k[a,\ b]$ is a set of continuous on $[a,\ b]$ functions that have continuous derivatives up to kth order.

In such specification of a problem, for given r it is necessary to find such grid Δ_r on the $[a,\ b]$ segment that a spline defined on it would give the optimal in a sense of least squares statistical estimation of vector A, which elements are assumed as ordinates in spline nodes. According to the least squares method [20, 21], it is necessary to reach the minimum of expression

$$(Y - XA) * (Y - XA) = \sum_{i=1}^{N} \left(y_i - \sum_{j=r}^{r} a_j x_{ij} \right)^2. \tag{2.65}$$

Vector A could be found from the equation

$$A = (X * X)^{-1} X * Y. \tag{2.66}$$

The confidence interval with probability level β for the estimations of A could be determined as

$$I_\beta^{(j)} = \left[a_j \mp \gamma_\beta \sqrt{\left\{(X * X)^{-1}\right\}_{jj} \frac{d}{N - r - 1}} \right], \tag{2.67}$$

where γ_β is a number found from the equation $\mathrm{P}\left\{|s_{N-r-1}| \le \gamma_\beta\right\} = 1 - \beta$ if random value s_{N-r-1} is distributed by Student law with $N - r - 1$ degrees of freedom; d is a sum of squares of the deviations of observations y_i from the spline values in corresponding points.

Equation (2.67) allows the determination of confidence intervals in every spline node and the confidence channel on the whole segment.

In order to obtain the forecast, an additional node is added to the set of spline nodes, which abscissa corresponds to the time of forecast. Using computer and enumerative technique, a grid was found satisfying the Eq. (2.66) and, at the same time, minimizing (2.65) on the set of all possible non-uniform grids and width of confidence interval in the forecast node.

As a result, the expected value and confidence interval of selected diagnostic parameter in the end-point of the forecast interval are found.

The frames of confidence channel are obtained by the linear interpolation of upper and lower limits of the confidence intervals in all nodes of the obtained spline, including forecasting node. An example of statistical spline is shown on Fig. 2.2.

Forecasting failure time of diagnostic constructive part is performed based on the selected diagnostic criterion. For example, the following rule could be used as a criterion: a node is considered to be defective when diagnostic parameter exceeds some predefined threshold.

For the above criterion, possible time t_f of the failure will be the intersection point of the obtained statistical spline with straight line, which ordinate is equal to predefined threshold (see Fig. 2.3).

Time interval, when the failure will happen with given confidence probability, is obtained in the intersection points of confidence channel with threshold linc. On Fig. 2.2 left and right limits are denoted as t_{f1} and t_{f2}.

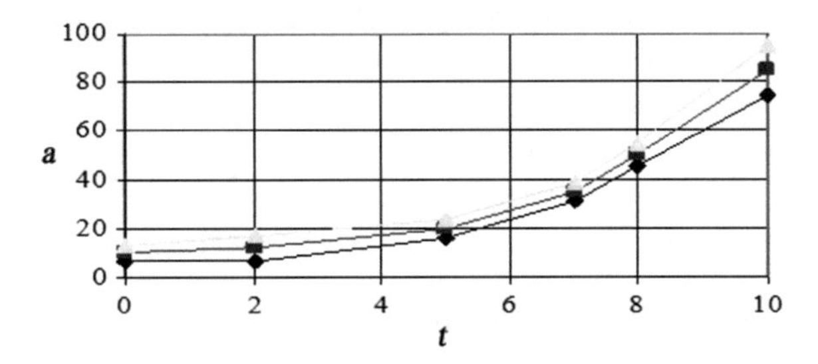

Fig. 2.2 An example of statistical spline with 6 nodes

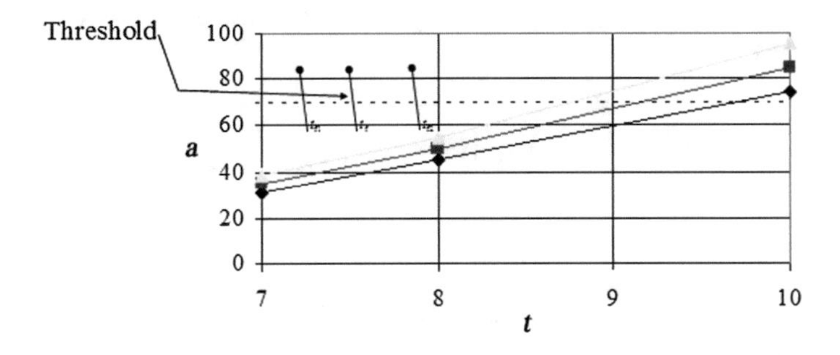

Fig. 2.3 Expected failure time t_f and corresponding confidence interval limits t_{f1}, t_{f2}

2.6.2 Forecasting Laminated Magnetic Cores Failures

To illustrate the method of electrical equipment failures time forecasting let's consider the following example.

In the Institute of Electrodynamics of National Academy of Science of Ukraine, the diagnostic expert system was created for shock vibration testing of laminated magnetic cores [20].

The model of laminated magnetic core was also created. It consists of the package of electric steel plates used in LP transformers. The pressing of plates could be adjusted with special pins.

A series of experiments were performed with this expert system. During the experiments, shock vibration wave was generated in the body of magnetic core. Vibration signals were measured by accelerometer. Then signals were amplified, converted to digital form by analog-to-digital converter and recorded to computer memory. Afterwards spectrograms of measured signals were obtained and the quantity of frequency peaks with amplitude exceeding the half of maximum value was counted. It was demonstrated that this diagnostic parameter could be successfully used for laminated magnetic cores diagnostics.

The corresponding diagnostic rule was formulated. In order to increase diagnostics reliability, several vibration signals are measured at the same time. Each of them is processed and the number of spectrogram frequency peaks is calculated. The average number of frequency peaks is used to make a decision about magnetic core technical status.

According to [20], a magnetic core is considered to be operable if the averaged quantity of frequency peaks does not exceed 6.255. Otherwise the plates' pressing is considered to be insufficient, which means that there is a magnetic core failure.

To verify proposed method an experiment was performed. During this experiment the plates pressing on the magnetic core model was little by little decreased. After every decrease, the diagnostic parameter (averaged quantity of frequency peaks in the spectrogram) was measured. Totally 12 measurements were performed.

The statistical spline was obtained as a result of observations processing. It is shown on Fig. 2.4. Since the deterioration of magnetic core was made artificially, the real time is not shown. Instead of it, the ordinal numbers of measurements are shown. They could be considered as conventional time units. Forecast interval is 5 time units, confidence probability is $p = 0.95$.

It is clear from the figure that continuing the experiments in the same conditions, pressing of magnetic core plates could reach low critical point at 13th measurement. The probability that magnetic core will "fail" between 13th and 16th measurement is 0.95.

Usage of statistical splines for electrical equipment failures forecasting allows obtaining time limits, in which the failure of equipment will happen with a predetermined probability level. This could be used for more accurate planning of maintenance and repair works terms. As a result, this allows complete utilization of equipment technological lifespan and increasing of its reliability.

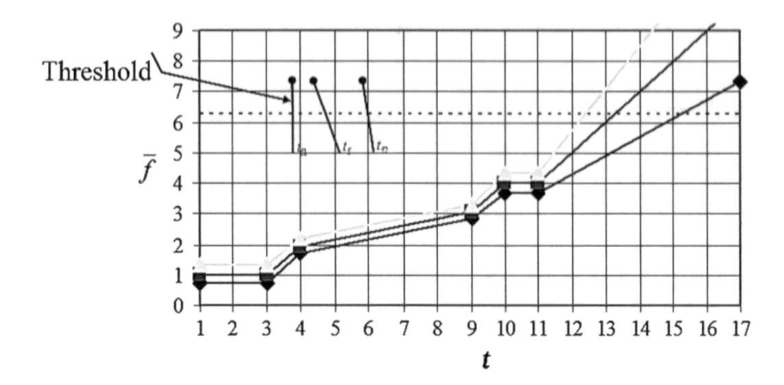

Fig. 2.4 Statistical spline representing the results of experiment with artificial deterioration of the laminated magnetic core model

2.6.3 Biomedical Applications

The Chernobyl accident resulted widespread contamination of environment. Many radionuclides were emitted to the environment from the reactor. Radionuclides entered human bodies mainly by inhalation, via drinking water and food-stuffs. Millions of people living around the Chernobyl zone including Kyiv city permanently accumulate radionuclides. ^{85}Sr and ^{137}Cs radioisotopes make the main contribution to the total radioactivity. These people need periodical decontaminating treatment to reduce negative consequences of internal irradiation.

The results of detailed investigations of radionuclide contamination in human bodies of inhabitants of Kiev after Chernobyl accident in 1986 have been discussed in the paper [21]. The investigations of ability of pectin—containing oral adsorbents to suppress ^{85}Sr and ^{137}Cs accumulation in Wistar rats have been studied by Ukrainian scientists [21]. These adsorbents could be used for decontaminating treatment of population living around the Chernobyl zone. One of the basic problems of decontaminating treatment of population is prediction of probable total radioactivity doses that were received by human bodies. The results of preliminary statistical investigations are commonly used to predict probable radioactivity doses. The statistical spline function method is proposed for prediction of confidence region for total radioactivity doses.

Analysis of curves that were represented in paper [21] shown that the experimental data of ^{85}Sr accumulation by rats met the above conditions. Therefore, the curves could be described by functional dependence (2.65) and the method of statistical spline functions could be applied for curve behavior prediction.

The channels of confidence intervals that correspond to confidence level $\mathrm{P} = 0.9$ and $A_{\mathrm{t}} = 5$ days array were constructed for the control group of rats and the group that received PVC adsorbent with application of experimental data and software MSAETS. The results are shown in Figs. 2.1 and 2.2. We are able to state that curves of radionuclide accumulation by the rats will be in these intervals in 90% of measurements.

The following fact could be a confirmation of the results obtained. We constructed confidence intervals for prediction of radionuclide accumulation by rats of both groups for the 33rd investigated day with application of results of the 28 days observations. The activity of ^{85}Sr accumulation by rats is within calculated confidence interval in both cases.

Calculation results of prediction confidence intervals of the activity of ^{85}Sr accumulation by rats of both groups for the 38th day with application of results of 33 days observation are shown in Figs. 2.5 and 2.6.

Represented results allow concluding that statistical spline method could be applied for prediction of radionuclides accumulation by living organisms.

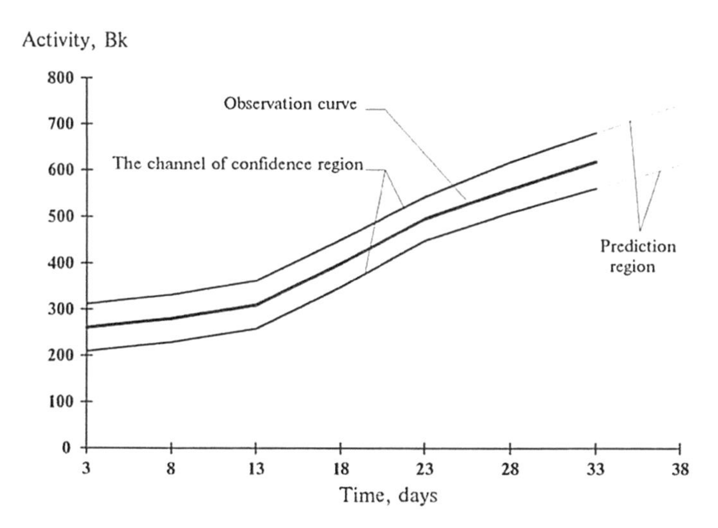

Fig. 2.5 Accumulation of ^{85}Sr by rats of control group

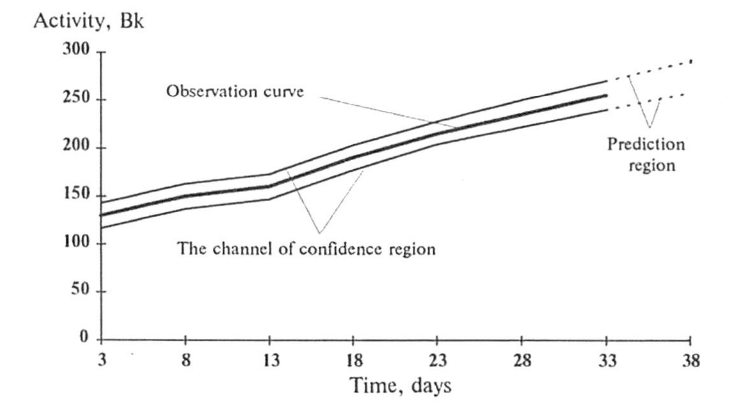

Fig. 2.6 Accumulation of ^{85}Sr by rats of PVC group

The method is attractive for solution of some problems of biophysics, radiobiology, biochemistry.

2.7 Estimation of Random Signals Stationarity

Experimental authors' studies have shown, however, that information signals could not always be assumed to be a stationary process. If this is not taken into account and vibration non-stationarity will entail a reduction in vibration diagnostics effectiveness and in isolated cases the diagnostics system to lose working capability. Note that some properties of stationary random processes were considered in [6].

Electric engine diagnostics could not always be automated. It is therefore necessary to introduce a preliminary diagnostics engine that is technically simpler to realize this could be referred to as malfunction detection. Here diagnostics-system regime produces information about the presence or absence of a malfunction. Diagnosis of defects should only be carried out when the electric engine is not functioning properly.

Some results of the method were published in [6]. The problem also was considered in [22, 23].

The basic aim of the chapter is to develop methods for estimating the information signals stationarity on a basis of certain statistical criteria and development of method for detecting malfunctions of electric machines using proposed stationarity-evaluation techniques.

The approach is based on the following consideration. Carrying out vibration diagnosis it is assumed that the appearance of a defect in an electric engine will lead to a change in vibration characteristics. If the vibration was a stationary random process prior to failure appearance, then in course of a time interval that includes the instant of failure appearance the vibration could no longer be assumed to be stationary in general case. Application of techniques for evaluating vibration stationarity makes it possible to detect the appearance of a failure; this is equivalent to detection of electric engine failure.

We introduce some definitions. The random process $\{\xi(t), t \in (-\infty, \infty)\}$ is said to be stationary in a broad sense if the mathematical expectation $M(\xi_t) = m$ is constant and independent of time t and correlation function $R(t_1, t_2) = R(t_1 - t_2)$ depends solely on the difference $|t_1 - t_2|$. If the finite-dimensional distributions of the process of $\xi(t)$ are independent of time origin, then $\xi(t)$ is said to be stationary in a strong sense.

A stationary process with discrete time ξ_t, $t \in Z$ (where $Z = \{\ldots, -1, 0, 1, \ldots\}$ is a set of integers) with independent values ξ_{tk} and ξ_{tl}, $l \neq k$, is a discrete white noise. It is stationary in broad sense if its mathematical expectation and variance are constant. The process ξ_t is stationary in narrow sense if its one-dimensional distribution function $F(x)$ is independent of time. To be rigorous we should assume the process ξ_t to be ergodic and non-periodically correlated.

Statistical analysis of the process ξ_{kt} is based on samples. If an experiment yields a series of numerical values $x_i = \xi_{ti}$, $i = \overline{1, n}$ of the process ξ_{kt}, then the set $x_1, x_2, \ldots, x_n$ is said to be a sample of size n for a realization of the process ξ_{kt} of the segment $k = \overline{1, n}$. The samples of the process ξ_{kt} should be used to test the hypothesis H_s that the process is a stationary state process against the alternative, hypothesis H_n, that it is a non-stationary state process.

As follows from the definition of a stationary state process in a strong sense, testing the hypothesis H_s consists of checking to see that N independent samples obtained at different times, each of size n, have been taken from the same random processes ξ_{kt} with distribution $F(x)$. If $F(x)$ is continuous, then H_s could be tested on the basis of Kolmogorov-Smirnov criterion. For this purpose we define the statistic

$$D_{n,n}^{i,j} = \max_{-\infty<x<\infty} \left| F_{in}(x) - F_{jn}(x) \right|, \tag{2.68}$$

where $F_{in}(x)$ and $F_{jn}(x)$, $i, j = \overline{1, N}$, $j \neq i$ are empirical distributions of ith and jth samples respectively. If for the pair of samples

$$D_{n,n}^{i,j} > k_{1-\alpha}\sqrt{n/2}, \tag{2.69}$$

then accept H_s and otherwise accept H_n. The values of quantiles $k_{1-\alpha}$ of Kolmogorov distribution are found from the tables given in [24]. Parameter α is the level of test significance. It is equal to the probability that a stationary process will be assumed to be a non-stationary on the basis of two samples, i.e. when $N = 2$. Evaluating the stationary state of information signals, we calculate the statistics (2.68) obtained from the pairs of samples of information signals. If calculated statistics satisfy the inequality (2.69), information signals are a stationary process with significance level less than α.

The procedure for estimating the stationary state of information signals becomes simpler if signal ξ_{kt} is normally distributed. In such a case, estimating its stationary state consists of testing the hypothesis H_s that variance and mathematical expectation are constant. It is convenient to implement such a test on the basis of criterion t_1, Cochran criterion t_2 and two-sample criterion t_3. The variance is first tested for constancy, and only after the mathematical expectation is estimated.

Estimating the stationary state of normal information signals ξ_{kt} could be carried out in the following sequence: for each of N samples we calculate the estimates of mathematical expectation m_i and variance σ_i^2, using the formulas

$$m_i = \frac{1}{n}\sum_{k=1}^{n} x_{ki},\ \sigma_i^2 = \frac{1}{n-1}\sum_{k=1}^{n} (x_{ki} - m_i)^2,\ i = \overline{1, N}, \tag{2.70}$$

where x_{ki} is the kth reading of the ith sample.

Constant Variance Tests
For sample size $n_i \neq n_j, n_i = n_j$ and number of samples $N = 2$ we use the following statistic (F-criterion)

$$t_i = \sigma_1^2 / \sigma_2^2. \tag{2.71}$$

For sample size, $n_i = n_j = n$ and number of samples $N > 2$ we use the following statistic (Cochran criterion)

$$t_i = \max\left(\sigma_i^2 / \sum_{i=1}^{N} \sigma_i^2\right). \tag{2.72}$$

Constant Mathematical Expectation Test
For sample size $n_i \neq n_j, n_i = n_j$ and number of samples $N = 2$ we use the following statistic (two-sample criterion, variance of compared samples assumed to be equal):

$$t_3 = |m_i - m_j| / \left(\left((n_i - 1)\sigma_i^2 + (n_j - 1)\sigma_j^2\right)/(n_i + n_j - 2)\right)^{1/2} \times ((n_i n_j)/(n_i + n_j))^{1/2}. \tag{2.73}$$

For the criteria chosen on the basis of calculated m_i and σ_i^2, we use formulas (2.71)–(2.73) to calculate the values of statistics t_i, $i = 1, 2, 3$. On the basis of statistics t_i distribution the critical values $C_{i,N,n,\alpha}$ have been calculated and tabulated in [24] for these criteria. Critical values depend on the number of samples N and sample size n, as well as on the significance level α. The hypothesis H_s is accepted that the variance or mathematical expectation is constant, provided

$$t_i < C_{i,N,n,\alpha}, \tag{2.74}$$

for corresponding criterion.

If H_s is accepted for σ_i^2 and for m_i we then make a decision that the normal random information signal being analyzed is a stationary random process.

To test the normality of information signals the criterion proposed in [24] could be applied. This consists of testing the hypothesis H that skewness and kurtosis coefficients fall within critical limits for the process being analyzed. To do this, on the basis of significance level and sample size, we use tables [24] to find critical limits for the estimates of skewness g_1 and kurtosis b_2 coefficients. If calculated g_1 and b_2 fall between the limits, we then accept hypothesis H. This means that the analyzed sample has been taken from a normal process, with an accuracy of up to the first four moments. The direct utilization of the samples obtained to estimate information signals stationarity by means of the proposed criteria involves errors, since in general case the readings obtained following quantification of information signals will be dependent. If we equate the amounts of information contained in samples obtained from the processes with dependent and independent readings, then the sample from the process with independent readings will be smaller in size. We refer to this size as the effective size n^*.

We consider an example of the method of stationary state estimation on the basis of roller bearings vibration signals studies [6]. It follows from the model proposed in [5–7] for EM roller bearings vibration and experimental data that the value of vibration autocorrelation function is equal to zero for a certain time interval Δt between readings. Readings of stationary normal vibration signals taken at intervals Δt will be independent, i.e. information signals could be represented as a random process $\xi_{k\Delta t}$. If $\Delta t \geq T_d = 1/f_d$ a sample with independent readings will have the maximum possible effective size that could be estimated from the following formula:

$$n^* = \mathrm{mod}[n/f_d S_o], \tag{2.75}$$

where mod $[a/b]$—integer part of division a/b; $S_0 \leq \Delta t$ is the time interval between the absolute maximum and the moment at which the autocorrelation function first passes through zero; f_d is the sampling frequency.

For the particular case in which $\Delta t = S_0 = T_d$, we obtain $n^* = n$.

An Example of Statistical Procedure Application

We consider the test application for studying roller bearings vibration signals. To simplify the procedure of estimating vibration stationarity from data corresponding to different test regimes, we selected samples for which the hypothesis H that they belong to a normal process has been accepted. We took respective significance levels of 2% for the criteria used to test to see whether skewness and kurtosis coefficients are equal to zero. The hypothesis H is accepted if the estimates of skewness coefficient g_1 and kurtosis b_2 coefficient for $n^* = 1000$ satisfy the inequalities $|g_1| < 0.17$, $-0.3 < b_2 < 0.39$ and, correspondingly, if for $n^* = 800$ we have $|g_1| < 0.202$, $-0.35 < b_2 < 0.46$. The critical limits of inequality were found from [24]. The number of samples selected and the corresponding test regimes are indicated in Table 2.1.

We used (2.70) for computer calculation of the estimates of mathematical expectation m_i and variance σ_i^2. On the basis of these estimates we used Cochran test and two-sample test to calculate the values of statistics t_2 and t_3 for each of the test regimes. The results of the analysis are shown in Table 2.1. Here the largest of all the statistics t_3 is calculated for all possible pairs of samples for given test regime.

Table 2.1 Calculated $C_{i,N,n,\alpha}$ values for $\alpha = 0.015$

Bearing test regime	N	t_2	$C_{2,N,n^*;0.01}$	$C_{2,N,n^*;0.15}$	t_3	$C_{2,N,\infty;0.01}$	$C_{2,N,\infty;0.15}$	H
Functioning	8	0.1339	0.1402	0.13562	1.38	2.576	1.44	H_S
Misaligned	10	0.1074	0.1129	0.1091	1.4	2.576	1.44	H_S
Defect on inner race misaligned	9	0.1278	0.1251	0.1251	1.54	2.576	1.44	H_N

In the next stage, we determine the critical values $C_{i,N,n,\alpha}$. We have to select the value α. For a constant sample size, the smaller α, the lower the probability that a stationary process will be assumed to be a non-stationary process, but as α diminishes there is an increase in the probability of assuming a non-stationary process to be stationary. A significance level $0.01 \le \alpha \le 0.15$ is often selected. We selected a significance level $\alpha = 0.015$. For the Cochran criteria, $C_{i,N,n,\alpha}$ values for $\alpha = 0.015$ were calculated for each of the test regimes on the basis of approximation formula proposed in [25]; they are shown in Table 2.1.

Comparing the values given in Table 2.1 for statistics t_2, t_3 and corresponding critical values shows that in accordance with the inequality (2.74), the hypothesis H_S, when variance and mathematical expectation are constant, is accepted with significance level of $\alpha = 0.015$ for each vibration of properly functioning bearings and misaligned bearings when ESh-176 lubricant is used. The hypothesis H_S that the variance is constant, and for the given case the hypothesis that vibrations are stationary for a bearing with a defect on the inner race, ESh-176 lubricant being used, are rejected with significance level $\alpha \ge 0.015$.

It follows from the table given in [25] for critical values $C_{i,N,n,\alpha}$ of two-sample criterion that for sample sizes $n = \infty$ and $n = 500$ these critical values differ by fractions of a percent if $0.01 \le \alpha \le 0.15$. As an example, $C_{2,2,\infty;0.01}/C_{2,2,300;0.01} = 1.0039$, $C_{2,2,\infty;0.1}/C_{2,2,300;0.1} = 1.0018$. Thus Table 2.1 shows $C_{2,2,\infty,\alpha}$ rather than $C_{2,2,1100,\alpha}$ for $\alpha = 0.01$ and $\alpha = 0.15$. Here both $C_{2,N,n^*,\alpha}$ and $C_{2,2,1100,\alpha}$ are monotonically decreasing and N, n^*—constant.

Comparison of the values given in Table 2.1 for statistics t_2, t_3 and corresponding critical values shows that, according to the inequality (2.74), the hypotheses H_S, when variance and mathematical expectation are constant, is accepted with significance level of $\alpha \le 0.15$ each for vibration of properly functioning bearing, misaligned bearings, and bearings with increased misalignment when ESh-176 lubricant being used. As a consequence, for each of these test regimes vibration could be assumed to be a stationary process. The hypothesis H_S that the variance is constant, and for the case of hypothesis that vibration is stationary for misaligned bearing with de: on the inner race, ESh-176 lubricant being used are rejected with significance level $\alpha \ge 0.01$. This means that if we choose a significance level $0.01 \le \alpha \le 0.15$, vibration-stationarity hypotheses will be accepted in the first three regimes and rejected in the last two.

Detection of Rolling Bearing Malfunctions

When a malfunction (certain types of defects) pears in an electric engine the vibration characteristics will change. This make it possible to suggest that over the course of the time interval that includes the instant of appearance of a malfunction the vibration will be a nonstationary process. Such an assumption could be verified practically by means of the methods developed for evaluation stationarity.

To illustrate this, let us look at the appearance of misalignment in a bearing using ESh-176 lubricant. Prior to its appearance, the vibration is a stationary process (Table 2.1). We now evaluate the vibration stationarity before and after misalignment appearance, analyzing 14 samples for this purpose (8 samples of vibration for the properly functioning bearing and 6 for the misaligned bearing). The statistic $t_2 = 0.0817$ for these samples while the critical values of Cochran test are $C_{2,N,4,1100:0.01} = 0.0808$, $C_{2,14,1100:0.15} = 0.0783$—(Table 2.2).

Comparing t_2 with the critical values we could see that for a significance level $\alpha \geq 0.01$ the hypothesis H_S that the vibration is stationary will be rejected. The appearance of misalignment in a bearing therefore led to a significant change in vibration parameters as a result over the course of time interval including the instant appearance of the misalignment the vibration cannot be assumed to be a stationary random process. It must be emphasized that this result was obtained under condition that the vibration of properly functioning or misaligned bearing has been assumed to be a stationary process according to the data of Table 2.1.

Analogously calculations of statistics t_2 and critical values were made other pairs of EM bearing test regimes. The results are shown in Table 2.2. For all pairs of regimes except for the "misaligned—misalignment increased" regime the hypothesis H_S that the variance is constant, and, therefore, that the vibration is stationary, is rejected with significance level $\alpha \geq 0.01$. For the "misaligned—misalignment increased" regime the hypothesis H_S that the variance is constant is accepted with significance level $\alpha \leq 0.15$. For this regime, however, the hypothesis H_S that mathematical expectation is constant is rejected with significance level $\alpha \geq 0.01$, so that vibration in "misaligned—misalignment increased" regime is also a nonstationary process.

Utilization of the suggested techniques for evaluating stationarity according to the method indicated therefore makes it possible to detect a change in vibration characteristics caused by the appearance of a defect or misalignment of bearings. In other words, it is possible to detect the malfunction of an electric engine.

It should be noted that the indicated method assumes stationarity of vibration in the initial regime. If this condition is not satisfied it is then necessary to carry out preliminary processing of vibration. In many cases such processing consists of eliminating the trend, and this could be done on the basis of techniques given in [6] as well as by bandpass filtering. Vibration stationarity could be evaluated follow-preliminary processes by means of the techniques described above.

If the vibration distribution is unknown it is necessary to use the Kolmogorov-Smirnov test to evaluate stationarity and detect malfunctions.

The existing algorithm is based on measurement of electric engine vibration velocity and its comparison with the maximum permissible effective value of vibration velocity. It is assumed that if the mum is exceeded this means that a defect has appeared. In addition to its positive characteristics, the existing algorithm gives rough estimates. As an example, it neglects the initial level (following fabrication) of vibration that differs even for machines of same type from the same series. Cases could occur in practice where a defect that occurs will increase the vibration level without exceeding the maximum. The corresponding algorithm for the first stage of vibration diagnosis fails to detect such feet, in contrast to the proposed malfunction-detection method.

A second drawback of the existing algorithm lies in the fact that to determine whether an EM is functioning properly we actually use a single vibration parameter,

Table 2.2 Calculated $C_{i,N,n,\alpha}$ values for $\alpha = 0.015$ and $\alpha \geq 0.01$

Pairs of bearing test regime (ESh-176 lubricant)	N	t_2	$C_{2,N,n^*;0.01}$	$C_{2,N,n^*;0.15}$	t_3	$C_{2,N,\infty;0.01}$	$C_{2,N,\infty;0.15}$	H
Functioning-misaligned	14	0.0817	0.0808	0.07833	3.75	2.576	1.44	H_N
Misaligned-misaligned increased	16	0.0683	0.0755	0.0732	3.45	2.576	1.44	H_N
Functioning-misaligned increased	18	0.0632	0.0629	0.06118	2.00	2.576	1.44	H_N
Functioning -defect on inner ring misaligned	17	0.0709	0.06853	0.06478	1.38	2.576	1.44	H_N
Misaligned-defect on inner ring misaligned	15	0.0756	0.0755	0.0732	3.46	2.576	1.44	H_N

the variance. Even in the case of normal vibration, the value of mathematical expectation also carries information, however. If the vibration is not a normal random process then information about the technical status is carried by higher-order moments. The proposed malfunction-detection method makes it possible to use nearly all of the information available in a one-dimensional vibration distribution function. Moreover, it is based not on measurement of the effective value of vibration velocity and its comparison with the maximum value, but on comparison of statistical characteristics of the vibration (vibration velocity or vibration acceleration), obtained at different instants in time. As an example, for the particular case of normal vibration if the ratio of two estimates of the variance (power) of vibration, calculated on the basis of two samples obtained at different times, exceeds the critical value then the proposed method indicates that over the time between two samples there was a malfunction in the electric engine.

2.8 A Procedure of Decision-Making Rule Development

Information on the technical or control status, obtained on the basis of statistical estimates of the informative parameters, is represented in the form of a random vector with coordinates $\Xi = a_1, a_2$.

Solution of the problem is reduced to accepting one of the following hypotheses:

1. H_1—random vector Ξ has distribution density of probabilities P_1;
2. H_2—random vector Ξ has distribution density of probabilities P_2.

Initially, we verified the hypotheses H_1, H_2 and then H_2 and H_3. Since the components of diagnostic parameters vector have a normal (Gaussian) distribution with a probability of 0.95, the combined distribution densities of diagnostic parameters have the form [12, 26]

$$P = \frac{1}{\sqrt{2\pi\,|M_m|}} \exp\left\{-\left[(\Xi - \Theta_m)^T M_m^{-1} (\Xi - \Theta_m)\right]\right\}, \qquad (2.76)$$

where Θ_m—is the vector of mathematical expectations $\Theta_m = \{\theta_{m1}, \theta_{m2}\}$; M_m is the covariance matrix of diagnostic (or control) parameters; $|M_m|$ is the determinant of the covariance matrix; Ξ is the vector of diagnostic (control) parameters, m is the number of hypotheses; l is the number of diagnostic parameters.

Determination of $\Theta_m\, M_m$ for hypothesis H_1 and H_2 is carried out using teaching sets corresponding to different technical (control) statuses of diagnosed (controlled) section by means of the following expressions

$$\theta_{ml} = \frac{1}{n}\sum_{i=1}^{n} a_{iml},\ M_m = \begin{vmatrix} \gamma_{1,1,m} & \gamma_{1,2,m} \\ \gamma_{2,1,m} & \gamma_{2,2,\mathrm{m}} \end{vmatrix},\ \gamma_{1,1,m} = \frac{1}{n-1}\sum_{j=1}^{n} (a_{1,j} - \theta_1)^2,$$

$$\gamma_{2,2,m} = \frac{1}{n-1}\sum_{j=1}^{n}\left(a_{2,j} - \theta_2\right)^2,\ \gamma_{2,1,m} = \frac{1}{n-1}\sum_{j=1}^{n}\left(a_{1,j} - \theta_1\right)\left(a_{2,j} - \theta_2\right),\ \gamma_{1,2,m} = \gamma_{2,1,m}, \tag{2.77}$$

where n is the volume of teaching set; θ_{ml} are the mathematical expectations of diagnostic (control) parameters; $\gamma_{1,1,m}$, $\gamma_{2,2,m}$ are the variance of diagnostic parameters; $\gamma_{1,2,m}$ is the mixed central moment of diagnostic parameters.

A similar procedure could be used to determine the elements of diagnostic spaces in different dimensions. The Neiman-Pearson procedure is based on the analysis of logarithm of the relationship of probability v and then is reduced to the optimum selection of some threshold c that separates the set of permissible values of u into two non-intersecting subsets, i.e. on the set of permissible values of v, it is necessary to select threshold c for which it could be concluded, at the given value of error of the first kind α, the fixed volume of sample n and the lowest value of error of the second type β, that the hypothesis H_1 holds if $v \geq c$ and H_2 if $v < c$.

The logarithm of probability ratio for the hypothesis H_1, H_2 is

$$u_N = \ln\sqrt{\frac{|M_2|}{|M_1|}} - \frac{(X - \Theta_1)^T M_1^{-1}(X - \Theta_1)}{2} + \frac{\left[(X - \Theta_2)^T M_2^{-1}(X - \Theta_2)\right]}{2}, \tag{2.78}$$

where Θ_m, M_m are respectively the vectors of mathematical expectations and the covariant matrices of diagnostic (control) parameters for the hypothesis H_m; X is the realization of observation vector Ξ; $(.)^T$—sign of conjugation; $(.)^{-1}$—sign of matrix inversion.

It is assumed that analyzed data of information signals is independent. If we carry out registration and analysis of N independent data from information (control) signals of sections, the logarithm of probability ratio of is $v = u_1 + u_2 + \cdots + u_n$. To determine threshold c and the required number of observations n, it is necessary to know the distribution of elementary probability u logarithm. This problem could be solved by several methods. We shall test some of them.

The distribution of elementary probability ratio logarithm could be estimated using Monte Carlo method (i.e. by carrying out statistical simulations). However, this method has a significant disadvantage—for every test condition of diagnosed (control) section it is necessary to carry out a relatively complex process of statistical simulations of elementary probability ratio logarithm and estimation of its distribution and to know in advance several parameters of such models.

There is another method for solving problems of this type. Using the transformation of elementary probability ratio logarithm parameters Θ_m, M_m and also realizations of observation vector X could be reduced to the distribution of a known type, for example, normal, χ^2. Transformations could be both linear and non-linear. Usage of these transformations of probability ratio logarithm parameter does not require complex process of statistical simulation.

Initially, it is convenient to reduce covariance matrixes M_m to diagonal type. For this purpose, it is necessary to determine the corresponding matrixes B_m for which the following relationships are hold:

$$B_m^T \times M_m \times B_m = A_m,\ A = \begin{bmatrix} \lambda_{1m} & 0 \\ 0 & \lambda_{2m} \end{bmatrix}. \tag{2.79}$$

Consequently $Y_m = B_m \times X$, $W_m = B_m \times \Theta_m$, where $\lambda_{1m}, \lambda_{2m}$ are the eigen numbers of the covariance matrix; B_m are matrixes of orthogonal transformations.

If we normalize the vector Y_m, W_m the matrixes A_1^{-1}, A_2^{-1} in the expression for elementary probability ratio logarithm are converted into unit matrixes.

Now, carrying out the transformation described previously, we are coming to the case of recognition of images of two classes of two-dimensional random quantities having normal distributions, identical diagonal matrixes and different vectors of mathematical expectations. Therefore, the elementary probability ratio logarithm has a normal distribution. According to the classic Neuman-Pearson procedure, specifying the probabilities of errors α and β, we could determine required number of observations N and threshold c using the expressions

$$N = \frac{\left(u_\alpha + u_\beta\right)^2 \times \sigma^2}{\left[M_1(u_n) - M_2(u_n)\right]^2}, \tag{2.80}$$

$$c = \frac{\left[M_1(u_n) + M_2(u_n)\right]}{2} + \frac{\sqrt{\sigma} \times \left(u_\alpha + u_\beta\right)}{2\sqrt{N}}, \tag{2.81}$$

where u_α, u_β are the quantiles of normal distribution; $M_1(u_n)$ is the value of mathematical expectation of elementary probability ratio logarithm for the hypothesis H_1; $M_2(u_n)$ is the value of mathematical expectation of elementary probability ratio for hypothesis H_2.

Consequently, the decision-making rule has the following form:

- if $v \geq c$, then the hypothesis H_1 is accepted with the probabilities of error of type I α and type II β;
- if $v < c$, then the hypothesis H_2 is accepted with the probabilities α of errors of type I and β of errors of type II.

Another linear transformation could be applied. If we use a preliminary centering of diagnostic control parameters and further we carry out orthogonal transformation of correlation matrixes of the parameters, then elementary probability ratio u has a non-centra χ^2 distribution with the number of degrees of freedom equals to the number of diagnostic parameters (dimension of diagnostic or control space).

Example of Decision Making Rules Application

Operation reliability improvement of electric engines and mechanisms is associated unavoidably with well-timed diagnostics of their technical status. Vibrodiagnostics method is the most promising one in a group of methods for diagnostics of technical status of electric engines and mechanisms elements.

In statistical approach, the solution of vibrodiagnostics problem usually consists of the following stages: construction of a mathematical model of machines and mechanisms diagnosed elements vibrations; verification of correspondence between mathematical model and experimental data: separation of the most informative diagnostic features using experimental data; formation of teaching sets of specimen corresponding to different technical states of the diagnosed machines of mechanisms; construction of decision making rules.

Vibrodiagnostics method is selected at the stage of constructing the mathematical model. Usually, vibrations are modeled using mathematical model with continuous parameters. However, the majority of modern devices and IMS in vibrodiagnostics include microprocessors, microcomputers and personal computers which functional algorithms are based on digital methods of information signals processing.

Taking this into account, vibrations should be described using mathematical models with discrete time.

In this work, we propose to use as mathematical models of machines and mechanisms elements vibrations one insufficiently studied group of random processes with discrete time—linear random autoregressive processes. We present the properties of linear autoregressive processes that could be used to solve other problems of engineering and could also be used in other areas of science.

AR models are used efficiently in problems of time series analysis, especially in problems of separating and classifying the information signals and also determining the response of a linear signal to a random effect.

There is a specific relationship between AR models and linear processes that makes it possible to separate a class of linear autoregressive processes.

Considered in [5, 6, 20], vibrodiagnostics methods contain the following three stages: "teaching" , "experiment design", "diagnostics". We will examine in detail the special features of using the method of decision rules making to diagnose 309ES h_2 bearings in a stand while analyzing rolling bearings vibrations. In "teaching" regime after estimating the stationarity of 309ES h_2 bearing vibrations data and carrying out autoregression analysis of vibration data [6], it is necessary to formulate teaching sets and estimate matrixes Θ_m, M_m corresponding to different technical statuses of studied bearings.

Representation (2.23) is the most common representation of autoregressive processes and makes it possible to describe a wide spectrum of information signals, including vibrations signals of electric engines elements.

We will analyze this assertion. The majority of studies concerned examination of properties and applications of autoregressive models for describing the physical processes assume that generating process is a Gaussian process. However, the experimental investigations show that vibrations of EM elements could not always be regarded as Gaussian random processes [27, 28].

The use of linear autoregressive models makes it possible to solve the problems of vibration signals classification within the framework of characteristic functions and corresponding distribution functions.

Experimental results show that parameters and autoregressive order could be used as possible diagnostic features for technical status of machines and mechanisms elements. Currently available methods of evaluating the autoregressive parameters could be divided into the following groups: the methods based on maximum probability estimates, the least squares method, and robust methods [6]. It should be mentioned that in case of Gaussian generating process ς_t, the estimates of maximum probability and the estimates of least squares differ only slightly, especially for the case of a large sample volume n from the realization of vibration process (n > 100).

The Yule-Walker method is one of the possible methods for evaluating the autoregressive parameters. The estimates of autoregressive parameters in this case are obtained by solving a system of the following equations:

$$\sum_{j=0}^{p} a_j r_{s-j} = 0,\ s = 1,\ 2, \ldots, \tag{2.82}$$

where r_s are the readings of correlation function of the process ξ_t; $a_1, \ldots, a_p$, are the autoregressive parameters; p is the autoregressive order.

If the length n of the process realization is relatively large, the true values of readings of correlation function $r(s)$ that are usually unknown could be replaced by the estimates.

If the vibration of a diagnosed element could be regarded as a stationary random process, the effective solution of equations system (2.84) is obtained using Levinson-Darbin algorithm [6].

Another important problem of vibrations autoregressive analysis is the estimate of autoregressive order. Autoregressive order could be estimated using various criteria, mainly AIC, FPE, BIC criteria proposed by Akaike, CAT criteria proposed by Parzen and HQ criteria introduced by Hannan and Quinn [6]. As indicated by the results of studies of autoregressive order statistical estimates using various criteria, the criteria most suitable for problems of vibrations autoregressive analysis is the Hannan-Quinn criteria that takes a strictly independent estimate of autoregressive order. The criteria is based on selecting the value of autoregressive order model that minimizes the expression

$$HQ(p) = \ln \hat{\sigma}_a^2 + 2pc\ \ln(\ln(n)), \tag{2.83}$$

where $\sigma_p^2 = \frac{1}{n}\sum_{k=1}^{p} a_k r(k)$, $c > 2$, $a_1, \ldots, a_p$ are autoregressive parameters; $r(k)$ are the counts of autocorrelation function of the process ξ_t; p is the autoregressive order; n is the volume of sample from vibration process realization.

The methods of estimating the autoregressive parameters described previously were used as a basis for algorithms of autoregressive prototype functioning of IMS for vibrodiagnostics developed at the Institute of Electrodynamics of the National Academy of Sciences of Ukraine. The prototype includes accelerometers, the unit for preliminary processing and filtration of vibrations, an analogue-digital convertor and a computing unit based on microcomputer. The IMS prototype for vibrodiagnostics

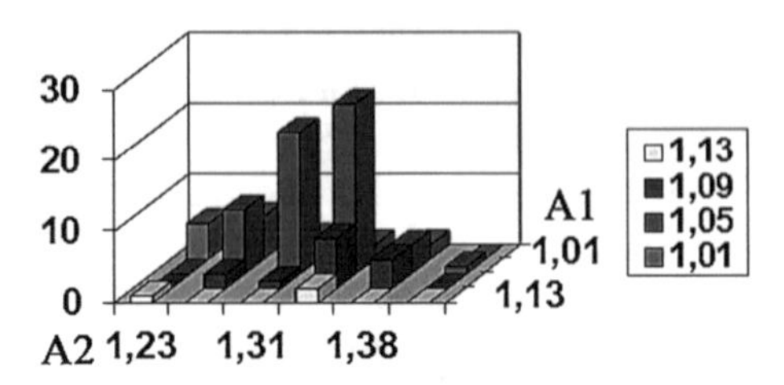

Fig. 2.7 Two-dimensional empirical histogram (bearing suitable for service)

was used in studying the statistical characteristics of vibrations of 309ESh2 rolling bearings positioned in a stand for examining vibrations of rolling bearings developed at the Institute of Electrodynamics. The vibrations of both repaired bearings and of bearings having a defect of "misalignment" type were studied.

At the initial stage, stationary samples were selected from realizations of vibrations for different technical stages of rolling bearings. The results of autoregressive analysis of more than 100 selected and classified as stationary realization vibrations, N – 2500 (N is the volume of the sample) in the frequency band 2–4 kHz, quantization frequency 16 kHz, were used to formulate the teaching sets for autoregressive coefficients $a_1, \ldots, a_p$ corresponding to different technical states of rolling bearings. Studies of autoregressive coefficients statistical characteristics make it possible to conclude that with the significance level of P – 0.05 the coefficients $a_1,\ a_2$ for these states of bearings: "in good working order", "misaligned", "greater misalignment", have a Gaussian distribution. Study of empirical histograms of distribution of autoregressive coefficients a_3 of working bearings show that these histograms are smoothed out by the first type of distribution in accordance with the system of Pearson curves [6].

The two-dimensional empirical histograms of distributions of autoregressive coefficients $a_1,\ a_2$ are shown in Fig. 2.7.

The procedures described previously were applied in the "teaching" regime.

The statistical estimates of autoregressive parameters, obtained as a result of autoregressive analysis of vibrations realization, represent the initial data for constructing the decision making rules.

We shall examine the construction of decision making rules using the Neuman-Pearson criterion on an example of diagnostics of misalignment and the degree of misalignment of 309ESh2 bearing positioned in a stand for examining the vibrations of rolling bearings.

Information on the technical status of diagnosed section, obtained on the basis of statistical estimates of diagnostic features is represented in the form of a random vector with coordinates $\Xi = a_1, a_2$.

The solution of this task is reduced to accepting one of the hypotheses: H_1—random vector Ξ has the distribution density of probabilities P_1 (bearings in good working order); H_2—the random vector Ξ has the density of distribution of probabilities P_2 (misaligned bearing); we verified the hypotheses H_1, H_2.

Using the relationships (2.79), we obtain matrixes corresponding to bearings suitable for service:

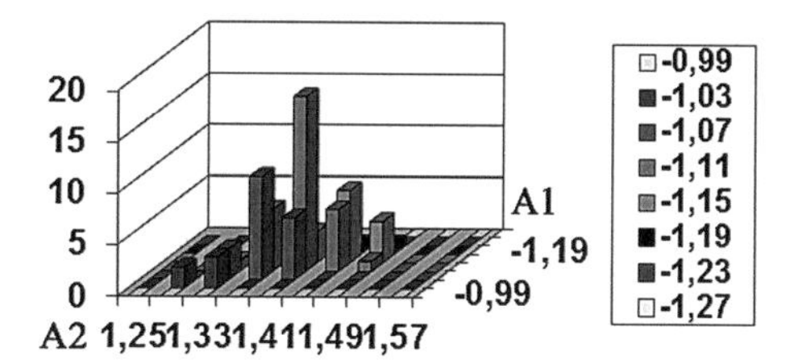

Fig. 2.8 Two-dimensional empirical histogram (misaligned bearing)

$$\Theta_1 = \begin{vmatrix} -1.0288 \\ 1.244 \end{vmatrix}, M_1 = \begin{vmatrix} 1.0162 \times 10^{-3} & -0.5207 \times 10^{-3} \\ -0.5207 \times 10^{-3} & 4.6817 \times 10^{-3} \end{vmatrix},$$

and for misaligned bearings

$$\Theta_2 = \begin{vmatrix} -1.1087 \\ 1.399 \end{vmatrix}, M_2 = \begin{vmatrix} 1.4517 \times 10^{-3} & -0.2549 \times 10^{-3} \\ -0.2549 \times 10^{-3} & 1.8445 \times 10^{-3} \end{vmatrix}.$$

It is now necessary to find matrixes of orthogonal transformations. For the example given, these matrices are (Fig. 2.8)

$$\mathbf{B_1} = \begin{vmatrix} 0.990437 & -0.137964 \\ 0.137964 & 0.990437 \end{vmatrix}, \mathbf{B_2} = \begin{vmatrix} 0.897313 & -0.441395 \\ 0.441395 & 0.897313 \end{vmatrix},$$

Expanding the matrix relationship $Y_m = B_m \times X$, we obtain the following equations for transition to statistics $\tilde{y}_1 \ldots \tilde{y}_4$:

$$\begin{aligned} &\tilde{y}_1 = 0,990437\tilde{a}_1 - 0,137964\tilde{a}_2, \ \tilde{y}_2 = 0,137964\tilde{a}_1 + 0,990437\tilde{a}_2, \\ &\tilde{y}_3 = 0,897313\tilde{a}_1 - 0,441395\tilde{a}_2, \ \tilde{y}_4 = 0,441395\tilde{a}_1 + 0,897313\tilde{a}_2. \end{aligned} \quad (2.84)$$

The transition to normalized statistics is carried out as follows:

$$\begin{aligned} &\mathbf{W_{1(n)}} = \begin{vmatrix} \mu_{11}/\sqrt{\lambda_{11}} \\ \mu_{12}/\sqrt{\lambda_{12}} \end{vmatrix} = \begin{vmatrix} -1.19055/\sqrt{0.949 * 10^{-3}} \\ 1.0902/\sqrt{4.754 * 10^{-3}} \end{vmatrix} = \begin{bmatrix} -38.642 \\ 15.811 \end{bmatrix}, \\ &\mathbf{W_{2(n)}} = \begin{vmatrix} \mu_{21}/\sqrt{\lambda_{21}} \\ \mu_{22}/\sqrt{\lambda_{22}} \end{vmatrix} = \begin{vmatrix} -1.6124/\sqrt{1.326 * 10^{-3}} \\ 0.7629/\sqrt{1.970 * 10^{-3}} \end{vmatrix} = \begin{bmatrix} -44.27 \\ 17.256 \end{bmatrix}, \\ &\mathbf{Y_{1(n)}} = \begin{vmatrix} \tilde{y}_1/\sqrt{\lambda_{11}} \\ \tilde{y}_2/\sqrt{\lambda_{12}} \end{vmatrix} = \begin{vmatrix} \tilde{y}_1/\sqrt{0.949 \times 10^{-3}} \\ \tilde{y}_2/\sqrt{4.754 \times 10^{-3}} \end{vmatrix}, \mathbf{Y_{2(n)}} = \begin{vmatrix} \tilde{y}_3/\sqrt{1.326 \times 10^{-3}} \\ \tilde{y}_4/\sqrt{1.970 \times 10^{-3}} \end{vmatrix}. \end{aligned} \quad (2.85)$$

To verify the hypotheses H_1, H_2, the expression for probability elementary ratio has the form:

$$u_{(1,2)n} = -\left[\boldsymbol{W}'_{1(\boldsymbol{n})} \times \boldsymbol{Y}_{1(\boldsymbol{n})} - \boldsymbol{W}'_{2(\boldsymbol{n})} \boldsymbol{Y}_{2(\boldsymbol{n})} + \frac{1}{2} \left(\boldsymbol{W}'_{2(\boldsymbol{n})} \times \boldsymbol{W}_{2(\boldsymbol{n})} - \boldsymbol{W}'_{1(\boldsymbol{n})} \times \boldsymbol{W}_{1(\boldsymbol{n})} \right) \right] =$$
$$= 1254.531 \times \tilde{y}_1 - 229.32 \times \tilde{y}_2 - 1215.988 \times \tilde{y}_3 + 388.426 \times \tilde{y}_4 - 257.648, \quad (2.86)$$

In the "experiment design" state, it is necessary to estimate the mathematical expectations and variance of elementary probability ratio logarithms for the hypotheses H_1, H_2. It is also necessary to determine the required number of observations n and the threshold c for given values of probabilities of errors of types I and II. For this example, the mathematical expectations of elementary probability ratio and also its variance for verification hypotheses H_1, H_2 are respectively: $M_1\left(u_{1,2}\right) = -163.302$, $M_2\left(u_{1,2}\right) = -202.567$, $\sigma = 151.756$. Taking into account the relationships (2.80), (2.81) and specifying probabilities of errors $\alpha = 0.05$, $\beta = 0.01$, we can determine the required number of observations n and threshold c:

$$N = \frac{\left(u_\alpha + u_\beta\right)^2 \times \sigma^2}{\left[M_1(u_n) - M_2(u_n)\right]^2} = \frac{(1.645+2.326)^2 \times 151.756}{(-163.302+202.567)^2} = 1.6 \approx 2,$$
$$- = \frac{\left[M_1(u_n) + M_2(u_n)\right]}{2} + \frac{\sqrt{\sigma^2} \times \left(u_\alpha + u_\beta\right)}{2\sqrt{N}} = \frac{-163.302+202.567}{2} + \frac{\sqrt{151.756}(2.326-1.645)}{2\sqrt{N}} = -180.935.$$

Consequently, the hypothesis H_1 (bearing is in good working order) is accepted if

$$v_{(1,2)n} \geq c_{1,2}. \quad (2.87)$$

where $v_{(1,2)n} = \left(u_{(1,2)I} + u_{(1,2)II}\right)/2$.

If the condition (2.87) is not fulfilled, we accept the hypothesis H_2 (bearing is misaligned).

In the "diagnostics" regime, technical status of examined bearing is tested as follows. According to the result obtained at the "experiment design" stage, to determine the misalignment of the 309ESh2 bearing with errors $\alpha = 0.05$, $\beta = 0.01$ placed on a stand for examining vibrations, it is necessary to record two independent data points of the vibrations and carry out their autoregressive analysis.

Let us assume that on the basis of vibrations autoregressive analysis results we have obtained the following estimates of the autoregressive coefficients:

$$\tilde{a}_{11} = -1.0733, \quad \tilde{a}_{12} = 1.351,$$
$$\tilde{a}_{21} = -1.0275, \quad \tilde{a}_{22} = 1.3508.$$

The first index of autoregressive coefficient denotes the number of data points of the vibration; the second is the order number of autoregressive coefficient. Using Eq. (2.86), we determine the values of the statistics $\tilde{y}_1 \ldots \tilde{y}_4$.

$$\tilde{y}_{11} = -1.24943, \quad \tilde{y}_{12} = 1.19, \quad \tilde{y}_{13} = -1.5594, \quad \tilde{y}_{14} = 0.73852,$$
$$\tilde{y}_{21} = -1.204, \quad \tilde{y}_{22} = 1.1961, \quad \tilde{y}_{23} = -1.51823, \quad \tilde{y}_{24} = 0.75856.$$

Substituting the resulting values of $\tilde{y}_1 \ldots \tilde{y}_4$ into the expressions of elementary probability ratio (2.87) we obtain

$$u_{(1,2)I} = -85.093, \; u_{(1,2)II} = -90.562,$$
$$v_{(1,2)n} = \left(u_{(1,2)I} + u_{(1,2)II}\right)/2 = -87.828$$

Since $v_{(1,2)} \geq c_{1,2}$ we accept the hypothesis H_1—the bearing is suitable for service.

References

1. Pugachev, V.S.: Probability Theory and Mathematical Statistics for Engineers (1984). ISBN 978-0-08-029148-2
2. Sinha, N.K., Telksnys, L.A.: Stochastic Control: Proceedings of the 2nd IFAC Symposium (1986). ISBN 978-0080334523
3. Zvaritch, V., Mislovitch, M., Martchenko, B.: White noise in information signal models. Appl. Math. Lett. **3**(7), 93–95 (1994). https://doi.org/10.1016/0893-9659(94)90120-1
4. Krasilnikov, A.I. Models of noise-type signals at the heat-and-power equipment diagnostic systems (2014)
5. Zvaritch, V., Glazkova, E.: Some singularities of kernels of linear AR and ARMA processes and their application to simulation of information signals. Comput. Prob. Electr. Eng. **1**(5), 71–74 (2015)
6. Capehart, B.L.: Information Technology for Energy Managers (2004). ISBN 978-0824746179
7. Marchenko, B., Zvaritch, V., Bedniy, N.: Linear random processes in some problems of information signal simulation. Electron Model **1**(23), 62–69 (2001)
8. Zvaritch, V.N., Marchenko, B.G.: Generating process characteristic function in the model of stationary linear AR-gamma process. Izvestiya Vysshikh Zavedenij Radioelectronika **8**(45), 12–18 (2002)
9. Zvaritch, V., Glazkova, E.: Application of linear AR and ARMA processes for simulation of power equipment diagnostic systems information signals. In: 2015 16th International Conference on Computational Problems of Electrical Engineering (CPEE), pp. 259–261 (2015). Lviv, Ukraine, Sept 2–5. https://doi.org/10.1109/cpee.2015.7333392
10. Zvaritch, V., Myslovitch, M., Martchenko, B.: The models of random periodic information signals on the white noise bases. Appl. Math. Lett. **3**(8), 87–89 (1995). https://doi.org/10.1016/0893-9659(95)00035-O
11. Javorskyj, L., Isayev, I., Majewski, J., Yuzefovych, R.: Component covariance analysis for periodically correlated random processes. Sig. Process. **4**(90), 1083–1102 (2010). https://doi.org/10.1016/j.sigpro.2009.07.031
12. Antoni, J., Guillet, F., Badaoui, M.E., Bonnardot, F.: Blind separation of convolved cyclostationary processes. Signal Process. **1**(85), 51–66 (2005). https://doi.org/10.1016/j.sigpro.2004.08.014
13. Hurd, H., Makagon, A., Miamee, A.G.: On AR(1) models with periodic and almost periodic coefficients. Stoch. Process. Appl. **1–2**(100), 167–185 (2002). https://doi.org/10.1016/S0304-4149(02)00094-7
14. Quinn, B.G.: Statistical methods of spectrum change detection. Digit. Signal Proc. **5**(16), 588–596 (2006). https://doi.org/10.1016/j.dsp.2004.12.011
15. Quinn, B.G.: Recent advances in rapid frequency estimation. Digit. Signal Proc. **6**(19), 942–948 (2009). https://doi.org/10.1016/j.dsp.2008.04.004
16. Nakamori, S.: Design of extended recursive Wiener fixed-point smoother and filter in discrete-time stochastic systems. Digit. Signal Proc. **1**(17), 360–370 (2007). https://doi.org/10.1016/j.dsp.2006.03.004

17. Labarre, D., Grivel, E., Berthoumieu, Y., Todini, E., Najim, M.: Consistent estimation of autoregressive parameters from noisy observations based on two interacting Kalman filters. Sig. Process. **10**(86), 2863–2876 (2006). https://doi.org/10.1016/j.sigpro.2005.12.001
18. Zvarich, V.N., Marchenko, B.G.: Linear autoregressive processes with periodic structures as models of information signals. Radioelectron. Commun. Syst. **7**(54), 367–372 (2011). https://doi.org/10.3103/S0735272711070041
19. Zvarich, V.N.: Peculiarities of finding characteristic functions of the generating process in the model of stationary linear AR(2) process with negative binomial distribution. Radioelectron. Commun. Syst. **12**(59), 567–573 (2016). https://doi.org/10.3103/S0735272716120050
20. Myslovich, M., Sysak, R., Khimjuk, I., Ulitko, O.: Forecasting of electrical equipment failures with usage of statistical spline-functions. In: 7th International Workshop "Computational Problems of Electrical Engineering" (2006)
21. Butsan, G.P.: Introduction to Probability Theory (2012). ISBN 978-966-360-209-7
22. Zhan, Y., Mechefske, C.K.: Robust detection of gearbox deterioration using compromised autoregressive modeling and Kolmogorov-Smirnov test statistic—Part I: Compromised autoregressive modeling with the aid of hypothesis tests and simulation analysis. Mech. Syst. Signal Process. **5**(21), 1953–1982 (2007). https://doi.org/10.1016/j.ymssp.2006.11.005
23. Zhan, Y., Mechefske, C.K.: Robust detection of gearbox deterioration using compromised autoregressive modeling and Kolmogorov–Smirnov test statistic. Part II: Experiment and application. Mech. Syst. Signal Process. **5**(21), 1983–2011 (2007). https://doi.org/10.1016/j.ymssp.2006.11.006
24. Bolshov, L.N., Smirnov, N.V.: Mathematical Statistics Tables (1983)
25. Kaźmierkowski M. P., Krishnan R., Blaabjerg F.: Control in power electronics: selected problems (2002)
26. Lopez, M.A.A., Flores, C.H., Garcia, E.G.: An intelligent tutoring system for turbine startup training of electrical power plant operators. Expert Syst. Appl. **1**(24), 95–101 (2003). https://doi.org/10.1016/S0957-4174(02)00087-8
27. Zvaritch, V.N., Malyarenko, A.P., Myslovitch, M.V., Martchenko, B.G.: Application of the statistical splines for prediction of radionuclide accumulation in living organisms. Fresenius Environ. Bull. **9**(3), 563–568 (1994)
28. Axelrod, A., Chowdhary, G.: The explore–exploit dilemma in nonstationary decision making under uncertainty. In: Busoniu L., Tamás L. (eds.) Handling Uncertainty and Networked Structure in Robot Control. Studies in Systems, Decision and Control, vol. 42, pp. 29–52 (2016). https://doi.org/10.1007/978-3-319-26327-4_2

Chapter 3
Simulation and Software for Diagnostic Systems

3.1 Computer Simulation of Noise and Rhythm Signals

Obtaining real information signals as a result of natural experiment is usually costly or even impossible [1–4]. In this regard, it is advisable to use computer simulation to solve problems of analyzing such signals. Let us consider methods and algorithms for simulation of noise and rhythm signals.

Simulation of noise signals. To simulate noise signals, we use their design models [5]. Diagnostic models. The basic model of noise signals are the processes of Bunimovich-Rice [5]:

$$\xi(t) = \sum_{k=1}^{\nu(t)} \eta_k h(t - t_k), \tag{3.1}$$

Status change of the diagnostic object leads to a change in process parameters (3.1), in particular, amplitudes distribution law η_k , pulses shape $h(t)$, and pulses appearance intensity λ. Therefore, diagnostic features are the change in probabilistic characteristics of noise signals.

Let us specify model parameters (3.1) and consider some typical shapes of elementary pulses (Table 3.1).

Since pulses amplitudes are positive, we give an exponential probability density of quantities η_k:

$$p_\eta(y) = \beta \exp(-\beta y) E(y), \beta > 0.$$

In Table 3.2 the cumulant coefficients $\gamma_s = \kappa_s \kappa_2^{-s/2}$, $s = \overline{3, 6}$, calculated in [2–4], of typical models of the process (3.1) for the values $\lambda\tau_0 = 1$ and $\lambda\tau_0 = 5$.

According to the Table 3.2 we can conclude that process distribution (3.1), even for a sufficiently large value $\lambda\tau_0 = 5$, differs significantly from the Gaussian distribution, for which $\gamma_s = 0$, $s \geq 3$ [6].

V. P. Babak et al., *Diagnostic Systems For Energy Equipments*, Studies in Systems, Decision and Control 281, https://doi.org/10.1007/978-3-030-44443-3_3

Table 3.1 The shape of elementary pulses

Number	Pulse shape	Analytical expression $h(t)$, $A > 0$, $\tau_0 > 0$
1	Rectangular	$h(t) = AE(t)E(\tau_0 - t)$
2	Saw-shaped	$h(t) = A(t/\tau_0)E(t)E(\tau_0 - t)$
3	Exponential-power	$h(t) = A(t/\tau_0)^b \exp(-t/\tau_0)E(t)$
4	Exponential-sine	$h(t) = A\exp(-t/\tau_0)\sin(\omega_0 t)E(t)$

Table 3.2 Cumulant coefficients of typical models

Pulses shapes	Parameter	Cumulative coefficients			
	$\lambda\tau_0$	γ_3	γ_4	γ_5	γ_6
1	1	2.121	6	21.213	90
	5	0.949	1.2	1.897	3.6
2	1	2.756	10.8	55.114	347.14
	5	1.232	2.16	4.93	13.886
3, $b = 0$	1	2	6	24	120
	5	0.894	1.2	2.147	4.8
3, $b = 1$	1	1.257	2.25	5.213	14.815
	5	0.562	0.45	0.466	0.593
4	1	0.357	8.978	11.234	297.132
	5	0.16	1.796	1.005	11.885

The simulation algorithm consists of the following stages [2–4]:

1. Determine sampling rate T_d of simulated signal by signal correlation interval ($T_d << \tau_{corr}$) or upper signal frequency ($T_d = 1/f_d, f_d >> f_в$).
2. Set the required duration of simulated realizations and form an array of discrete time points t for which signal samples will be modeled. The time interval between neighboring elements of the array is T_d.
3. Obtain an array of random moments t_k of elementary pulses appearance based on a given intensity λ of the Poisson events stream where the intervals between neighboring moments t_k are distributed according to the exponential law.
4. Form an array of random amplitudes η_k of pulses with the desired distribution law.
5. Form k-th elementary pulse. For this, a function is simulated $h(t)$, offset by a value t_k and multiplied by amplitude value η_k.
6. Simulate process realization (3.1) by summing obtained elementary pulses for all values from the time array.

Simulation results. Let us simulate noise signals realization on the computer for specific values of diagnostic models parameters.

Set the parameter values: $A = 1$, $\beta = 1$, $\tau_0 = 1$ μs. Choose the exponential-sinus pulses basic frequency $f_0 = \omega_0 / 2\pi = 5$ MHz. The study will be done for the intensity of $\lambda = 5 \times 10^6$ s^{-1} that corresponds to $\lambda\tau_0 = 5$. We choose sampling rate $T_d = 10$ ns (sampling frequency is 100 MHz) that is shorter than duration τ_0 of elementary pulse by 100 times. Let us set realizations duration equal to $T = 0.5$ ms. Then the sample size for each realization is $N = 5 \times 10^4$ samples. To ensure simulated process stationarity, time origin will be transferred to the point $t_0 = 20\tau_0$.

Figure 3.1 shows simulated realization of the process (3.1) with the shape of impulses 3 ($b = 0$) and 4.

To test the simulation algorithm effectiveness, we compare the values of theoretical and experimental initial moments. Theoretical moments α_s of the process (3.1) are found using formulas for the connection of initial moments with cumulants κ_s:

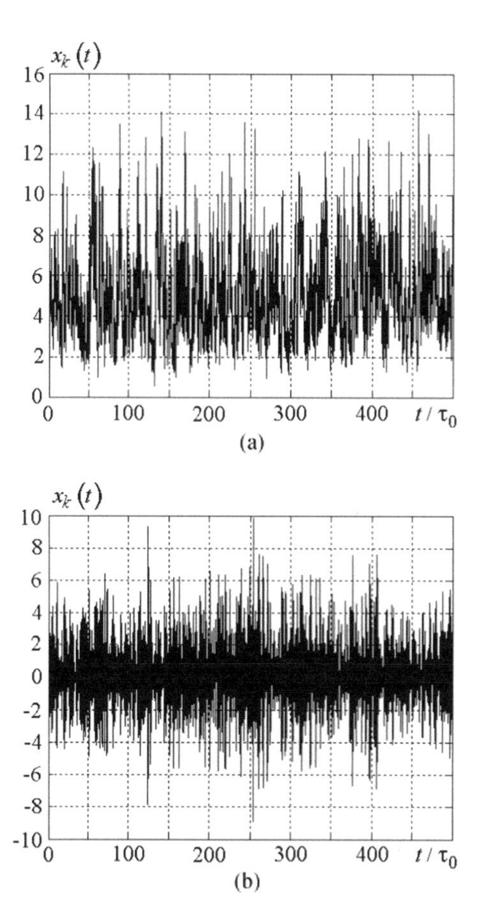

Fig. 3.1 Bunimovich-Rice process realization: **a** with exponential pulse shape 3; **b** with exponentially-sinus pulse shape 4

Table 3.3 Values of theoretical and experimental moments

Moments	Pulse shape				
	1	2	3, $b = 0$	3, $b = 1$	4
α_1	5	2.5	5	5	0.159
$\hat{\alpha}_{1\,cp}$	5.038	2.548	5.027	4.987	0.157
α_2	35	9.58	30	27.5	2,525
$\hat{\alpha}_{2\,cp}$	35.28	9.87	30.19	27.34	2.505
α_3	305	48.13	210	164.7	1.827
$\hat{\alpha}_{3\,cp}$	305.7	49.89	210.7	162.9	1.729
α_4	3145	296.4	1680	1066	30.73
$\hat{\alpha}_{4\,cp}$	3118.3	308.2	1672.2	1044	30.45

$$\alpha_1 = \kappa_1;\ \alpha_2 = \kappa_2 + \kappa_1^2;\ \alpha_3 = \kappa_3 + 3\kappa_1\kappa_2 + \kappa_1^3;$$

$$\alpha_4 = \kappa_4 + 4\kappa_1\kappa_3 + 3\kappa_2^2 + 6\kappa_2\kappa_1^2 + \kappa_1^4.$$

Table 3.3 shows theoretical values, obtained in [5], of the initial moments of studied models and their estimates $\hat{\alpha}_{s\,cp}$ obtained by averaging over 30 realizations.

According to Table 3.3, we could conclude that theoretical and experimental values of the process moments are well coordinated, since the relative error for this realizations ensemble did not exceed 5.4% (in the case of moment α_3 for pulse shape 4). Figure 3.2 shows the histograms and graphs of theoretical probability densities of the process (3.1) with pulse shapes 3 ($b = 0$) and 4, constructed using theoretical initial moments, rather than their estimates.

Charts in Fig. 3.2 show the consistency between theoretical density of probabilities and histograms.

Simulation of rhythmic signals. It is known that generalized models of rhythmic signals are stationary random processes with discrete spectrum and periodically correlated random processes [5].

Valid stationary random processes with discrete spectrum could be represented as follows:

$$\xi(t) = \sum_{k=-\infty}^{\infty} \gamma_k e^{i2\pi f_k t}, \tag{3.2}$$

where γ_k—independent equally distributed complex random variables, in which $\mathbf{M}[\gamma_k] = 0$, $\mathbf{D}[\gamma_k] = \sigma_k^2$, where γ_k and γ_{-k} form complex conjugate pairs, and frequencies $f_k = f_{-k}$, $f_0 = 0$.

Correlation function of the process (3.2) is almost a periodic function and is equal to $R(\tau) = \sigma_0^2 + 2\sum_{k=1}^{\infty} \sigma_k^2 \cos 2\pi f_k \tau$.

Let us note individual cases of the model (3.2).

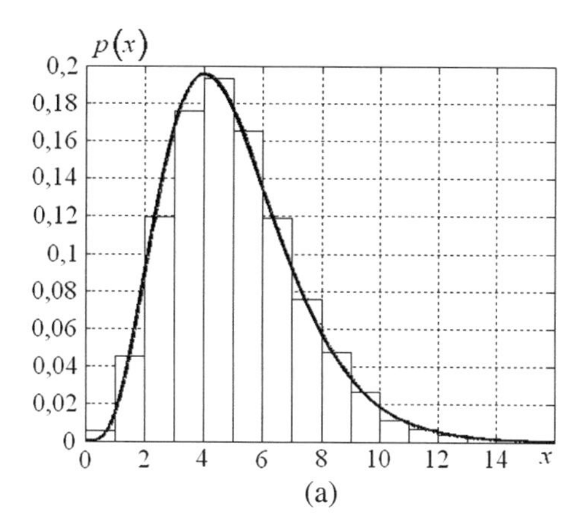

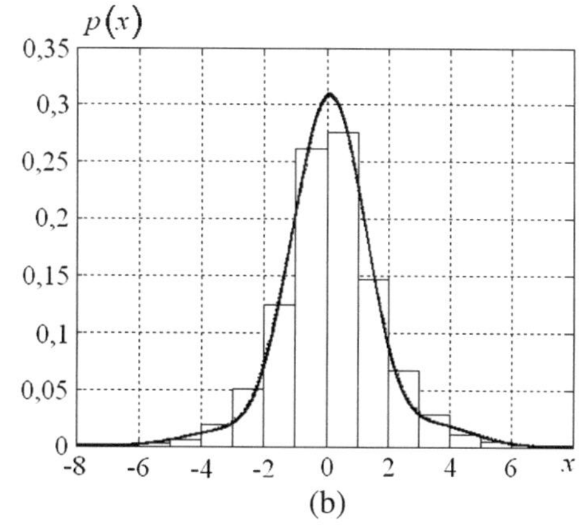

Fig. 3.2 Theoretical probability density and process histogram (3.1): **a** with exponential pulses shape 3; **b** with exponentially-sinus pulses shape 4

1. If frequencies f_k are multiple, then $f_k = kf_1$, correlation function is a periodic function with a period $T_0 > 0$, $R(\tau) = R(\tau + T_0)$ and

$$\sigma_k^2 = \frac{1}{T_0} \int_0^{T_0} R(\tau) e^{-i2\pi k f_1 \tau} d\tau.$$

2. Let γ_k—degenerate random variables equal to C_k. Then the model of rhythmic signals is deterministic almost periodic functions

$$\xi(t) = \sum_{k=-\infty}^{\infty} C_k e^{i2\pi f_k t}, \tag{3.3}$$

3. If in the formula (3.3) frequencies f_k are multiple, that is $f_k = kf_1$, C_k are Fourier coefficients, then the process $\xi(t)$ is a deterministic periodic function with a period $T_0 = 1/f_1$.

So, for simulation we will use formula (3.2), which we rewrite in actual form

$$\xi(t) = \gamma_0 + \sum_{k=1}^{\infty} \gamma_k \cos(2\pi f_k t - \varphi_k), \tag{3.4}$$

In simulation, number of terms in the formula (3.4) is finite.
Simulation algorithm contains the following steps:

1. Set the number of harmonics n.
2. Set the amplitude of harmonics γ_k.
3. Set the initial phases φ_k.
4. Set the realization duration T.
5. Determine the sampling rate based on the frequency of higher harmonics f_n:

$$T_d = 1/f_d, \ f_d >> f_n.$$

6. Simulate realization process, summing up all the harmonics.

Simulation results. We simulate two processes with discrete spectrum ($n = 10$); with multiple frequencies: $f_1 = 50$ Hz, $f_k = kf_1, k = \overline{1,\ 10}$; and with non-multiple frequencies: $f_1 = 50$ Hz, $f_2 = 75$ Hz, $f_3 = 130$ Hz, $f_4 = 175$ Hz, $f_5 = 225$ Hz, $f_6 = 275$ Hz, $f_7 = 310$ Hz, $f_8 = 360$ Hz, $f_9 = 410$ Hz, $f_{10} = 470$ Hz. We assign to both processes the following simulation parameters: $\gamma_0 = 0$, $\gamma_1 = 1$, $\gamma_2 = 0.9$, $\gamma_3 = 0.8$, $\gamma_4 = 0.7$, $\gamma_5 = 0.6$, $\gamma_6 = 0.5$, $\gamma_7 = 0.4$, $\gamma_8 = 0.3$, $\gamma_9 = 0.2$, $\gamma_{10} = 0.1$; random variable φ_k is evenly distributed over an interval $[0;\ 2\pi]$. We set implementation time to 10 s, and sampling rate is 20 kHz.

Figure 3.3 shows realization of simulated processes with discrete spectrum for the case of multiple and non-multiple frequencies; Fig. 3.4—estimates of their correlation functions, Fig. 3.5—estimates of spectral densities. Data window size in spectral analysis was 16.384 readings that determines resolution ratio at frequency of 20,000/16,384 = 1.22 Hz. It is seen from figures that in the case of multiple frequencies, the process and its correlation function are periodic functions, which is not the case with non-multiple frequencies; in both cases, spectral densities have discrete components at frequencies f_k.

Periodically correlated random processes with a period $T_0 > 0$ are non-stationary processes and satisfy the following conditions:

(1) $m(t + T_0) = m(t)$;
(2) $R(t_1 + T_0,\ t_2 + T_0) = R(t_1,\ t_2)$.

According to [6–8] periodically correlated random processes can be obtained by periodic repetition of stationary process segment T_0.

We use for the simulation a Gaussian stationary process with independent values with parameters $m = 0$ and $\sigma = 1$.

We set the following simulation parameters: realization time—100 s, sampling rate—20 kHz (sample size $N = 2 \times 10^6$ counts). Data window size for spectral

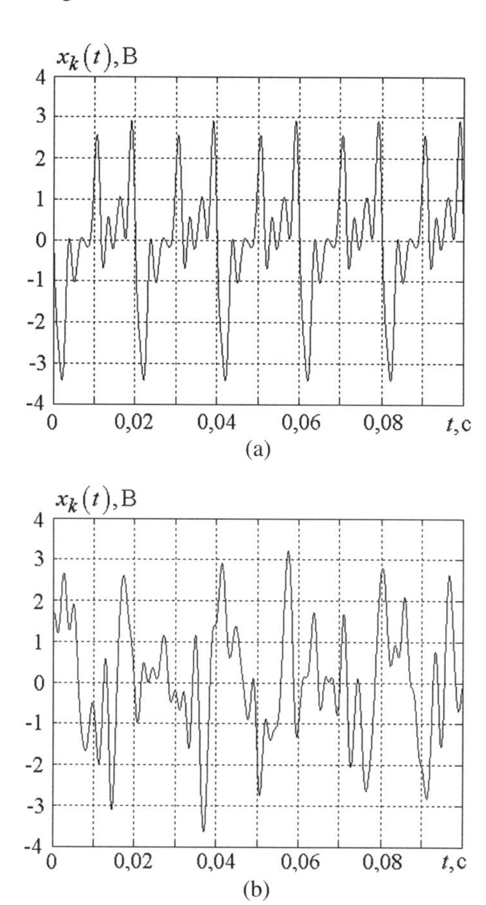

Fig. 3.3 Realization of processes with discrete spectrum: **a** multiple frequencies; **b** non-multiple frequencies

analysis is 8192, which determines the resolution at a frequency of 20,000/8192 = 2.44 Hz. We isolate the realization from initial process segment with length $T_0 = 0.02$ (400 counts) and repeat it 4999 times. The resulting realization has duration of 100 s (sample size $N = 2 \times 10^6$ counts). Note that selected repetition period corresponds to a frequency of 50 Hz that is rotary equipment operation characteristic.

Figure 3.6 shows realizations of simulated processes—stationary and periodically correlated, Fig. 3.7—their correlation functions estimates, Fig. 3.8—spectral densities estimates.

From Figs. 3.6, 3.7 and 3.8 implies that realizations of both signals are similar, but the correlation function of periodically correlated process is periodic with an interval of 0.02 s, and its spectral density instead of fluctuations around the average value of 5×10^{-5} V^2 s has discrete components at frequencies, multiple 50 Hz.

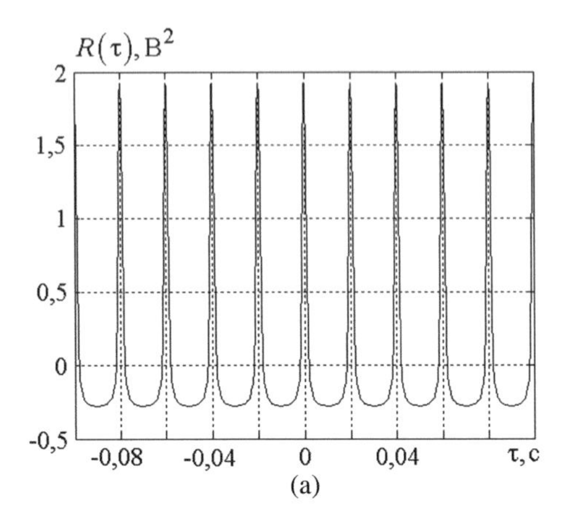

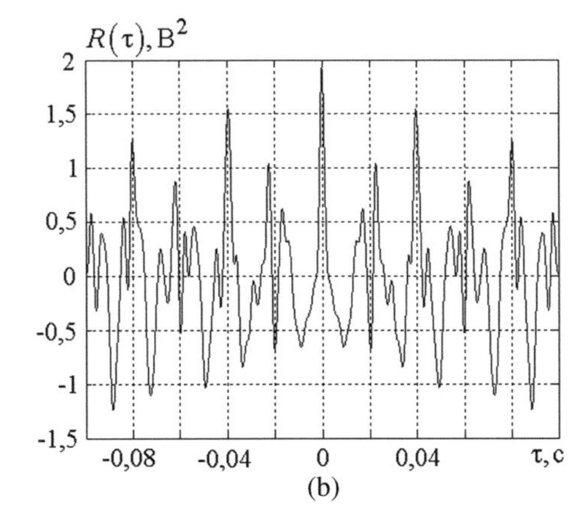

Fig. 3.4 Estimates of correlation functions of processes with discrete spectrum: **a** multiple frequencies; **b** non-multiple frequencies

3.2 Diagnostic Systems Software

Applied software is one of the main components of diagnostic systems; it implements the function of interaction with hardware system modules, managing information flows within the system, implementing algorithms for digital and statistical processing of information signals, building information fields images, forming the user interface, storing the information received and outcomes, etc.

The general structure of diagnostic systems software is presented in Fig. 3.9.

As a development platform, National Instrument' Lab View environment was used. This product is a powerful tool for creating its own software, focused on solving both research and application problems, and has a comprehensive library of embedded tools for digital signal processing and mathematical data processing.

The program module for interaction with modules of hardware system complex (control and measuring equipment modules) provides software coordination of these modules with the software system as a whole. To link measurement module software

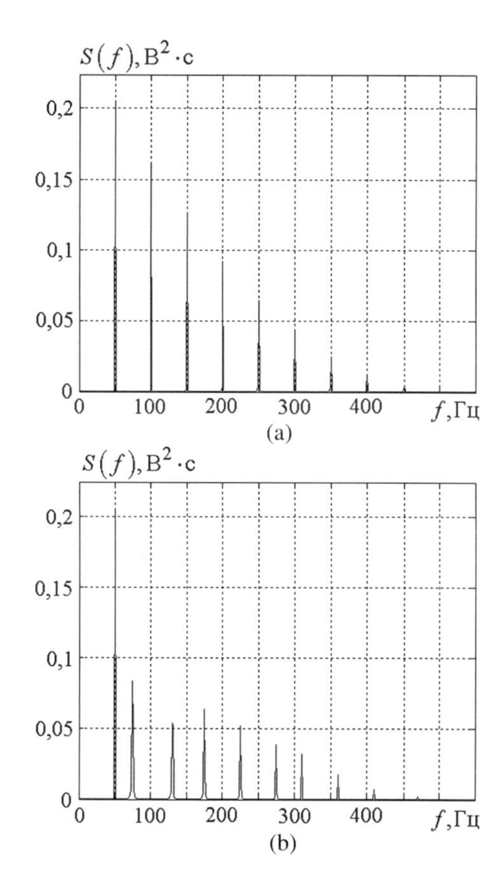

Fig. 3.5 Estimates of spectral densities of processes with discrete spectrum: **a** multiple frequencies; **b** non-multiple frequencies

to the system software a special library of programmer functions is used or API of functions—application programming interface, which is a set of tools for interacting with different types of software.

The module for interaction with system modules provides data exchange with the driver components that are responsible for the following operations: searching for connected device; getting device identifier; configuration and starting conversion process; checking the current status of the process of collecting and transmitting a signal about its completion; reading the received data and transfer it for further processing; stoping data collection and shutting down the device.

When system starts, shell library is loaded. With its help device working functions become available.

The first stage consists of procedures for searching and initializing respective control and measurement modules, after which they become ready for work as part of the monitoring system. Data collection starting signal is formed by calling a function that is responsible for setting working parameters: each individual module polling frequency, synchronization mode, data format, etc.

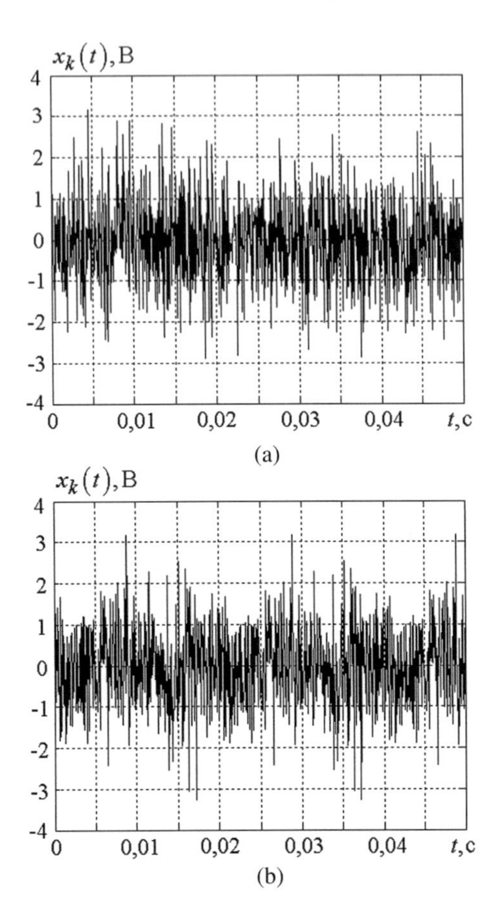

Fig. 3.6 Realizations of simulated processes: **a** stationary; **b** periodically correlated

After completion the assembly cycle, measurement data arrays, which are transferred to data transmission channel input, are formed in control modules buffers. It could be a communication interface. The resulting arrays are transmitted to software modules that realize information signals processing functions—modules for primary informative parameters separation, their statistical processing and control process management.

The module for primary informative parameters separation could also be a part of the system depending on modules structure of control and measuring equipment of the system. This module provides accumulation of primary data and manages their transmission through a specialized interface to other software system modules for further statistical processing, building information fields' distributions, and automatic recognition of data classes that correspond to different statuses of studied object.

Primary data accumulation module stores the input information arrays and reduces their dimension by selecting certain diagnostic features that are used for further

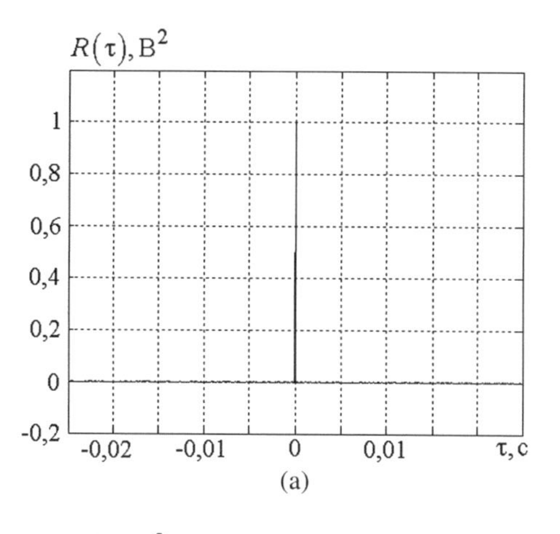

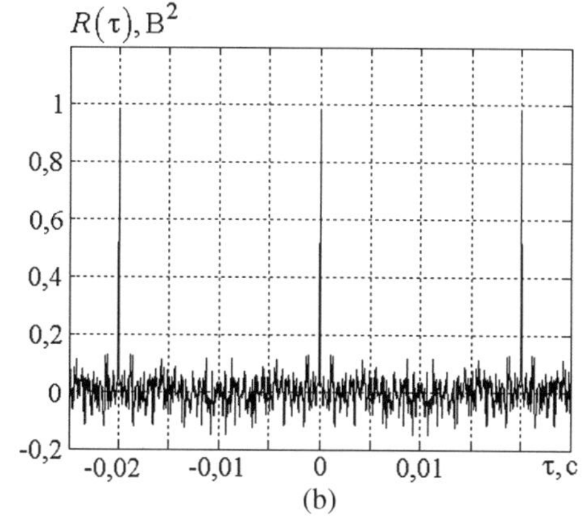

Fig. 3.7 Processes correlation functions estimates: **a** stationary; **b** periodically correlated

processing, which makes it possible to improve data processing and system operation as a whole, as well as normalization of input vector to bring it to a given values range.

Video and thermal imaging data processing module (if appropriate units are presented in system' hardware part) provides the following algorithms for images processing—increasing spatial resolution, increasing contrast, image filtering, separating geometric primitives, and binding to standards. In addition, this module allows geometric measurements using 1D and 2D images to localize objects coordinates and sizes, as well as brightness measurements in tasks of processing information coming from thermal cameras. Developing this module, IMAQ Vision library of Lab VIEW environment was used.

Data class recognition module carries out input data classification for certain given characteristics, splitting the set of attribute vectors into clusters, and recognizing different classes (images). This module is realized on the basis of modified neural network that allowed us to build decision-making rules based on the minimum

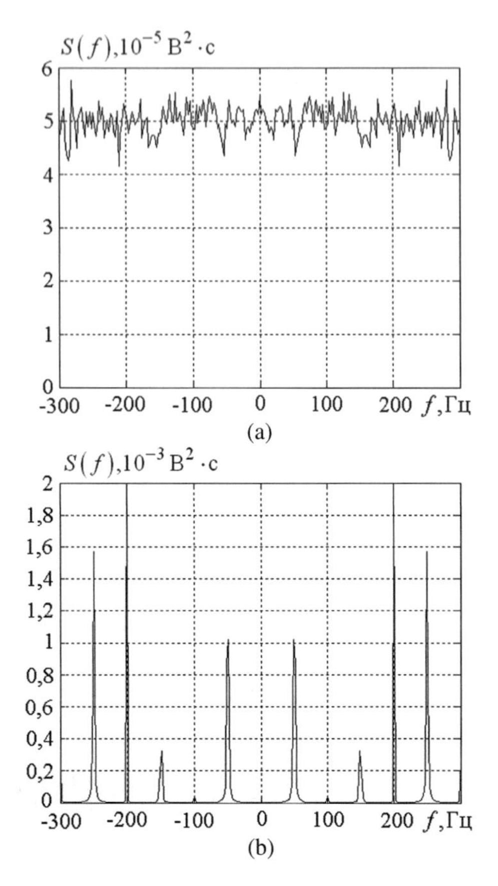

Fig. 3.8 Processes spectral density estimates: **a** stationary; **b** periodically correlated

possible initial information, as well as to dynamically modify these rules in diagnostic process. These characteristics were obtained by making changes to the classical structures of neural networks and creating new algorithms for their operation.

The functional load of statistical data processing unit is the most important in monitoring system, so we will consider its software modules in more detail. Software structure of this block is presented in Fig. 3.10.

The module for preliminary sample data censoring is intended primarily to reduce the effect of sample values with excessive errors and to identify progressive or periodic trends.

The first problem is solved by data filtration—window or median, depending on the input vector dimension, or using of statistical criteria, which, for a given confidence probability, allows calculation of boundary selective values (according to Romanovsky and Dixon criteria).

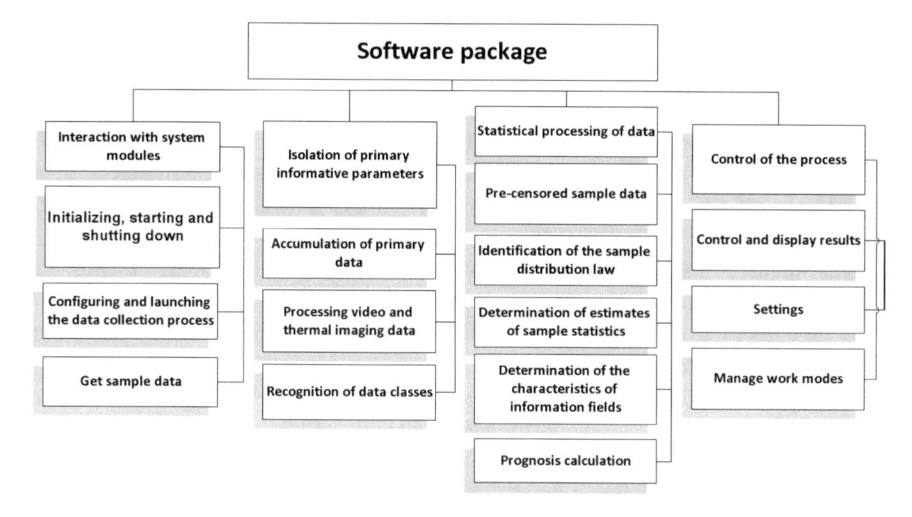

Fig. 3.9 Software structure

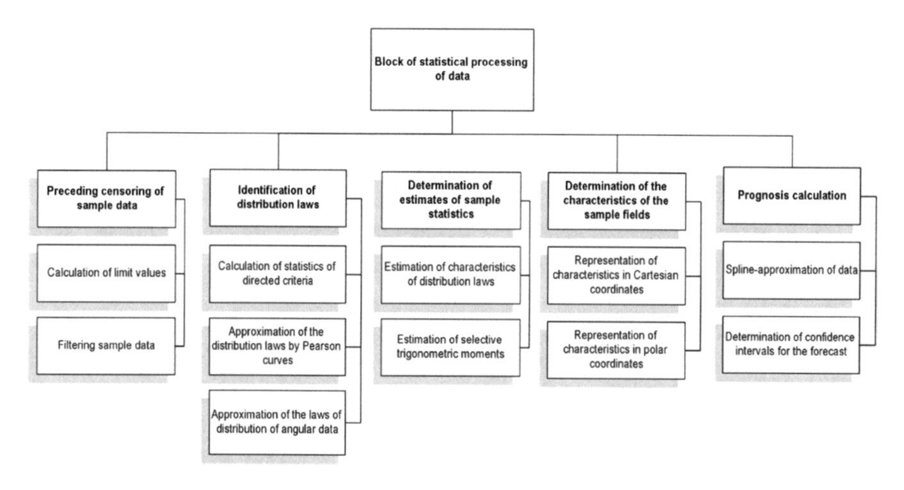

Fig. 3.10 Software structure of statistical data processing unit

To identify trends, we use the series criteria and the Foster-Stewart method that allows us to establish not only mathematical expectation trend, but also observations series variance trend.

Identification of experimental data distribution laws is necessary not only to confirm the fact of their homogeneity, but also to select one or another model of information field, and also could be used as a diagnostic feature.

Distribution laws identification module uses the following statistical tools: directed testing criteria for Gassiness (compiled criterion) and uniformity (Frosini criterion), approximation of linear data distribution laws by Pearson curves, and approximation of angular data distributions laws by Mises and wounded Gaussian distributions.

Distribution laws identification module works as follows: previously censored sampled data is divided into linear and angular; linear data is checked for the belonging of selective distribution law to Gaussian or uniform directed criteria. If these hypotheses are not confirmed, an approximation by Pearson distributions is made that is considered in subsequent calculations.

Angular data distribution laws are approximated either by Mises or wounded Gaussian distributions. The use of Mises distribution is more acceptable, since it has a mathematical record compared to the wounded Gaussian that leads to simpler estimates of distribution parameters.

Selective statistics determination module provides assessment of experimental data statistical parameters considering specified distribution law specified for further use in calculation of fields' characteristics or control object status prediction.

Information fields' characteristics determination is based on calculated statistical characteristics and considering selected field model.

Information fields could be constructed in both Cartesian and polar coordinates. Field characteristics definition module allows constructing field images in 3D as well as studying their characteristics changes over time.

Forecast calculation module performs regressive model construction of measured parameters behavior over time based on experimental data spline-approximation.

Developed forecasting technique allows obtaining forecast confidence intervals, where, with given probability, informative parameters values could be in the future. It actually allows predicting information fields' behavior over time.

User interface formation, necessary settings installation, operating modes selection and indication of results are assigned to control unit, which architecture is shown in Fig. 3.11.

Basic software code of developed software system is executed in modular structure that allows connecting and integrating of previously created subroutines and additional modules in high-level languages into the main program code, working with dynamic dll libraries, expanding the functionality of additional software modules and functions without making significant changes to basic software structure.

This approach allows adding necessary or eliminate unnecessary elements in system software without any complications, modernize and adapt system to the change of tasks and working conditions, etc. Getting a large amount of diagnostic information that needs processing complicates data processing algorithms and increases the time to analyze them. Therefore, at present, the use of intelligent computer technologies for solving monitoring objects status recognition problems is of high relevance. Recognition (clustering) problem is solved by developing and applying an artificial neural network.

The choice of this solution is due to neural networks ability to perform operations of processing, comparing and categorizing images that are not available to traditional mathematics, and possibility of self-learning and self-organization allows creating powerful intelligent systems to solve monitoring and diagnostics problems.

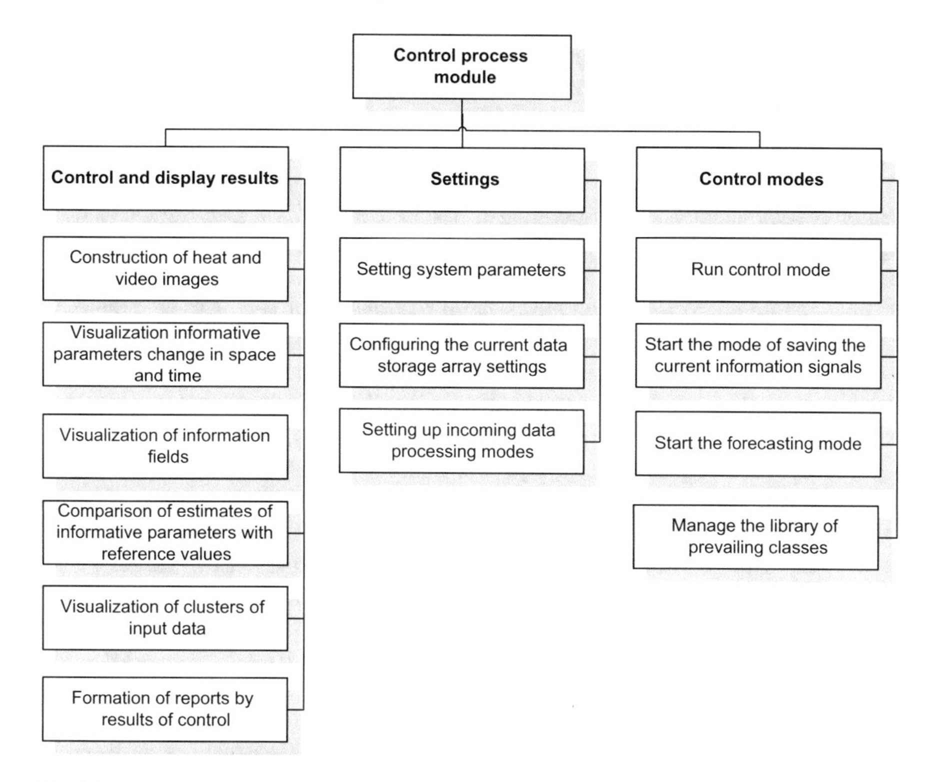

Fig. 3.11 Control process management module structure

3.3 Neural Networks in Diagnostic Systems

Neural network classifier in monitoring system provides nonlinear separation of features space, the ability to automatically replenish class library, high reliability of clustering in cases of minimal initial information about recognized images and limited number of images for learning [4, 9].

Solving the problem of clustering using neural networks there is a dilemma—it is necessary, during operation, to ensure, on the one hand, the plasticity of neural network memory (the ability to perceive new data and to create new clusters), and on the other hand to maintain stability, which guarantees that information about already known clusters is not destroyed and does not collapse [5]. This is achieved using the neural network of adaptive resonance theory—ART-2 [9].

ART-2 neural network is not sensitive to input vectors presentation order, could work both with binary, and with continuous signals, has high speed and provides high authenticity of input signals identification. In addition, ART-2 network has a feature of classification errors self-correcting after a certain number of learning samples repeat presentation [10].

The ART network [3] is a vector classifier. The input vector is classified depending on image similarity to previously recorded images by the network. The decision about input vector identification the ART network expresses in a form of excitation of one of the recognition layer neurons. If the input vector does not match any of "reference" images, a new category is created (a new neuron is allocated and a new vector is remembered), corresponding to input vector. If it is determined that the input vector is similar to one of the vectors (standards) met before, according to a certain similarity criterion, the reference vector in the neural network memory will be changed (studied) under the influence of the new input vector to become more similar to this input vector.

Remembered image will not change if the current input vector does not appear to be very similar to it. New image could create additional classification categories, but it will not be able to force the existing memory to change.

ART network consists of two layers of neurons F1 (layer comparison) and F2 (recognition layer) [6]. F1 layer contains n neural elements that correspond to the input image dimension. Each neuron has synoptic relationships with F2 layer elements. Each F2 layer neuron characterizes some images cluster. Each network layer corresponds to its own matrix of weight coefficients W and V. The direct weighted vector W_i corresponds to i cluster, the reciprocal weighted vector V_j characterizes integral image corresponding to this cluster. The activity of F1 and F2 layers neurons is responsible for short-term memory, and weight vectors W and V—for long-term memory.

Basic equations describing the ART-2 network F1 comparison layer operation:

$$p_i = u_i + \sum_j g(y_j) v_{j,i}, q_i = p_i / (e + |p|), u_i = z_i / (e + |z|),$$

$$z_i = f(t_i) + b \cdot f(q_i), s_i = x_i + a \cdot u_i, t_i = s_i / (e + |s|),$$

where y_j—the output of the j-th neuron of F_2 recognition layer; $v_{j,i}$—elements of weight coefficients V matrix; a and b—coefficients determined experimentally; e—parameter characterizing the relationship between F_1 and F_2 layers neurons operation time, $0 < e << 1$; $f(x)$—neurons activation nonlinear signaling function.

Neurons activation function could be either continuously differentiated

$$f(x) = \begin{cases} 2 \cdot \theta \cdot x^2 / (x^2 + \theta^2), & \text{at } 0 \le x < \theta \\ x, & \text{at } x \ge \theta \end{cases},$$

or piecewise linear:

$$f(x) = \begin{cases} 0, & \text{at } 0 \le x < \theta \\ x, & \text{at } x \ge \theta \end{cases}.$$

ART-2 F_2 recognition layer operation is described by the following equations:

$$T_j = \sum_i p_i w_{i,j},\ T_k = \max\{T_j : j = \overline{1, m}\},$$

$$g(y_k) = \begin{cases} d, \text{ at } T_k = \max\limits_j (T_j) \\ 0, \text{ otherwise} \end{cases},$$

where $w_{i,j}$—weight coefficients W matrix elements; d—a constant determined experimentally.

Thus, comparison device will receive vector p_i:

$$p_i = \begin{cases} u_i, & \text{with inactive neurons of the layer } F_2 \\ u_i + d \cdot v_{k,i}, & \text{with active neuron } k \end{cases}.$$

Comparison device activates mute signal if the following condition is not met:

$$\rho / (e + |r|) \geq 1,$$

where $\rho \in [0, 1]$—classifier sensitivity coefficient; $r = (r_1, r_2, \ldots, r_n)$—a vector characterizing the degree of difference of the input vector X from the "reference" vector W_k in network memory:

$$r_i = (u_i + c \cdot p_i) / (e + |u| + |c \cdot p|),$$

where c—weighted coefficient, selected from inequality:

$$c \leq (1 - d) / d.$$

In case of correct input vector classification, the mute signal is not activated, and weighted coefficients of W and V matrices are modified as follows:

$$v_{j,i}^{new} = v_{j,i}^{old} + \Delta v_{j,i},\ \Delta v_{j,i} = g(y_j)[p_i - v_{j,i}] = d \cdot (p_i - v_{k,i}),$$

$$w_{i,j}^{new} = w_{i,j}^{old} + \Delta w_{i,j},\ \Delta w_{j,i} = g(y_j)[p_i - w_{i,j}] = d \cdot (p_i - w_{i,k}).$$

where $v_{j,i}^{old}$, $v_{j,i}^{new}$, $w_{i,j}^{old}$, $w_{i,j}^{new}$—V and W matrices weighted coefficients before and after modification accordingly.

At the beginning of network operation and formation of a new neuron (in case of a new class formation), corresponding weighted coefficients values are initialized by the initial values:

$$v_{j,i} = 0,\ w_{i,j} \leq \frac{1}{(1 - d)\sqrt{N}},\ i = \overline{1, N}, \quad j = \overline{1, m},$$

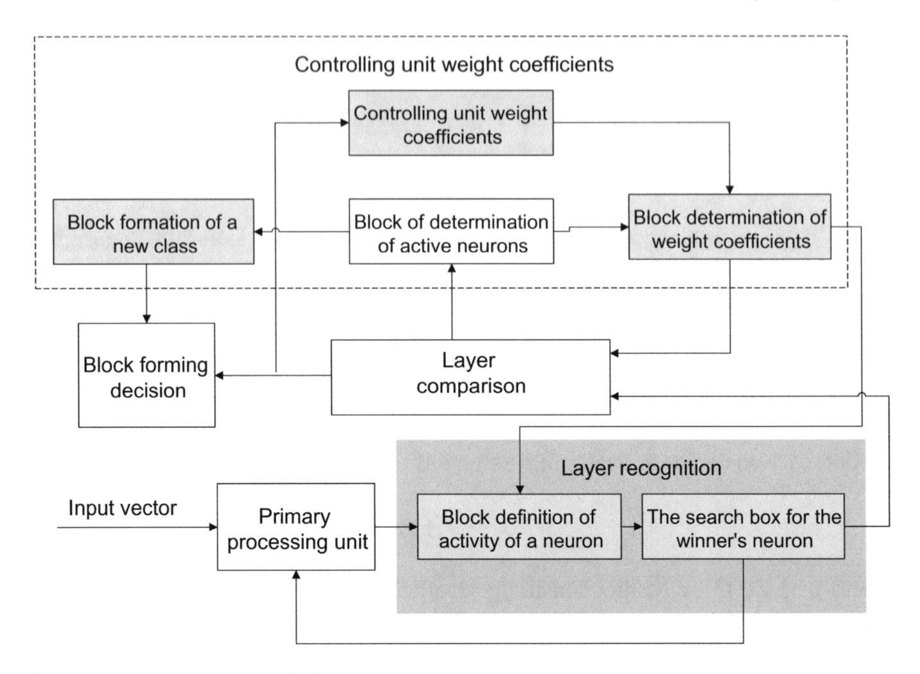

Fig. 3.12 Classifier structural diagram based on ART-2 neural network

where N—input vector of X dimension; m—number of neurons in F_2 recognition layer (number of classes in memory).

In order to increase efficiency of ART-2 network and clustering reliability, a new network architecture, algorithm for its operation and learning was developed [11].

Figure 3.12 shows the developed ART-2 network structural scheme, where blocks, modified and supplemented in relation to classical ART-2 network, are selected.

Let's analyze the differences of the new structure.

Firstly, in classical ART-2 network implementation, weighted coefficients of W and V matrices have almost identical values, that is, there is a duplication of values [12]. In developed ART-2 network together with two matrixes of weighted coefficients an operation algorithm is implemented that provides the use of one weighted coefficients matrix, which reduces system memory consumption and the number of computing operations (in the process of modifying the weighted coefficients during learning and addressing the elements of corresponding matrices in calculations).

Developed network has the following relationship:

$$V = W^T.$$

Thus, required memory size and number of operations while modifying the weighting factors during learning are reduced twice.

Secondly, in classical ART-2 network realization, de facto only one criterion for assessment of input vector belonging to a certain class is selected. In new developed

ART-2 neural network architecture, two weighted coefficients matrices were used to calculate two criteria for assessment of input vector belonging to a certain class, but their elements values are modified when learning in a way that differs from the classical theory.

In the proposed network, the algorithm of F_2 recognition layer is described by the equations:

$$T_j = \sum_i x_i w_{i,j},\ T_k = \max\{T_j : j = \overline{1,m}\},$$

where $w_{i,j}$—weighted coefficients W matrix elements; x_i – input vector X elements.

F_1 comparison layer operation remains unchanged. Modification of weighted coefficients during learning is carried out according to the following algorithm:

$$v_{j,i}^{new} = v_{j,i}^{old} + \Delta v_{j,i},\ \Delta v_{j,i} = g(y_j)[u_i - v_{j,i}] = d \cdot (u_i - v_{k,i}),$$

$$w_{i,j}^{new} = w_{i,j}^{old} + \Delta w_{i,j},\ \Delta w_{j,i} = g(y_j)[x_i - w_{i,j}] = d \cdot (x_i - w_{i,k}),$$

where $v_{j,i}^{old}$, $v_{j,i}^{new}$, $w_{i,j}^{old}$, $w_{i,j}^{new}$—V and W matrices weighted coefficients before and after modification.

All other calculations are performed according to classical ART-2 network algorithm. This classifier based on ART-2 network has the following advantages in identification problem solution: stored information stability, the ability to dynamically expand its own knowledge base, high impedance, invariance to input vectors presentation order, the ability to automatically correct mistakes that were obtained at previous stages of learning [12–14].

References

1. Inoue, H., Harrigan, J.J., Reid, S.R.: Review of inverse analysis for indirect measurement of impact force. Appl. Mech. Rev. **54**(6), 503–524 (2001). https://doi.org/10.1115/1.1420194
2. Yan, G., Zhou, L.: Impact load identification of composite structure using genetic algorithms. J. Sound Vib. **3–5**(319), 869–884 (2009). https://doi.org/10.1016/j.jsv.2008.06.051
3. Allen, M., Carne, T. Comparison of inverse structural filter (ISF) and sum of weighted accelerations (SWAT) time domain force identification methods. In: 47th AIAA/ASME/ASCE/AHS/ASC Structures, Structural Dynamics, and Materials Conference, Newport, Rhode Island, 1–4 May 2006. https://doi.org/10.2514/6.2006-1885
4. Bataineh, M., Marler, T.: Neural network for regression problems with reduced training sets. Neural Netw. **95**, 1–9 (2017). https://doi.org/10.1016/j.neunet.2017.07.018
5. Li, H., Li, C., Huang, T.: Periodicity and stability for variable-time impulsive neural networks. Neural Netw. **94**, 24–33 (2017). https://doi.org/10.1016/j.neunet.2017.06.006
6. Babak, V., Eremenko V., Zaporozhets, A.: Research of diagnostic parameters of composite materials using Johnson distribution. Int. J. Comput. **18**(4) (2019)

7. Krasilnikov, A., Beregun, V., Harmash, O.: Analysis of estimation errors of the fifth and sixth order cumulants. In: 2019 IEEE 39th International Conference on Electronics and Nanotechnology (ELNANO), April 16–18, 2019, Kyiv, Ukraine, pp. 754–759. https://doi.org/10.1109/elnano.2019.8783910
8. Babak, V., Filonenko, S., Kalita, V.: Model of acoustic emission signal at self-accelerated crack development. Aviation **9**(3), 3–8 (2005). https://doi.org/10.3846/16487788.2005.9635904
9. Babak, V., Filonenko, S., Kalita, V.: Acoustic emission under temperature tests of materials. Aviation **9**(4), 24–28 (2005). https://doi.org/10.3846/16487788.2005.9635914
10. Chen, C.H.: Ultrasonic and Advanced Methods for Nondestructive Testing and Material Characterization (2007). ISBN: 978-981-270-409-2
11. Eremenko, V., Zaporozhets, A., Isaenko, V., Babikova, K.: Application of wavelet transform for determining diagnostic signs. In: CEUR Workshop Proceedings, vol. 2387, pp. 202–214 (2019). http://ceur-ws.org/Vol-2387/20190202.pdf
12. Eremenko, V.S., Pereidenko, A.V., Roganov, V.O.: System of standardless diagnostic of cell panels based on Fuzzy-ART neural network. In: 2011 Microwaves, Radar and Remote Sensing Symposium, 25–27 Aug 2011, Kiev, Ukraine. https://doi.org/10.1109/mrrs.2011.6053630
13. Grosse, C.U., Ohtsu, M. Acoustic Emission Testing (2008). ISBN: 978-3-540-69972-9
14. Milovančević, M., Milenković, D., Troha, S.: The optimization of the vibrodiagnostic method applied on turbo machines. Trans. FAMENA **3**(33), 60–70 (2009)

Chapter 4
Technical Provision of Diagnostic Systems

4.1 Acoustic Devices and Systems for Power Plant Modules Diagnostics

Passive systems of functional diagnostics where noise and rhythmic signals are the source of information, arising from natural objects functioning [1], play an important role in controlling and diagnosing heat-energy equipment during their operation in the last decades.

Noise signals are the result of mechanical, aerodynamical, hydrodynamical and tribomechanical processes, which accompany heat and power equipment units operation, and manifest themselves in form of acoustic, magnetic, electric, thermal noise or broadband vibrations.

Rhythmic signals are the result of interaction of parts in the kinematic pairs of gas turbines, gas-piston engines, electric machines, compressors, etc., and, as a rule, appear in form of narrow-band multifrequency vibrations.

Passive diagnostic systems based on the use of noise and rhythm signals are widely used to control and determine the heat energy objects technical status [2].

The most common systems for diagnosing heat energy equipment that use noise signals—acoustic emission systems, acoustic contact—leak detection systems, and rhythmic signals—vibro-acoustic systems [1].

Acoustic-emission systems [2]. In solids under the influence of stresses created by an external load, the material structure is dynamically restructured at a microscopic or macroscopic levels. As a result, continuous and discrete acoustic emission (AE) occurred.

Continuous acoustic emission occurs during plastic deformation of bodies as a result of the motion of dislocations, twinning, and diffusionless phase transitions. Discrete emission results from microcracks formation. When material yield point is exceeded, it is a consequence of dynamic rearrangement of material structure under the action of high internal stresses caused by the accumulation of dislocations. The appearance of discrete emission characterizes the initial stage of destruction and it

V. P. Babak et al., *Diagnostic Systems For Energy Equipments*, Studies in Systems, Decision and Control 281, https://doi.org/10.1007/978-3-030-44443-3_4

is associated with the formation, development and spreading of cracks.

Acoustic-emission signals recorded during plastic deformation and crack growth are significantly different. During plastic deformation, various metals and alloys emit a large number of exponential pulses of AE of small amplitude. Acoustic-emission signals of this type are recorded as a continuous process, and their realizations are similar to the realization of electronic devices thermal noise.

During the crack developing, every crack jump generates a separate exponential pulse of a discrete high-amplitude of AE. The development of a crack occurs unevenly and leads to the formation of a discrete acoustic-emission signal in the form of a chaotic sequence of short individual pulses or partially overlapping pulses, which have a high energy level. The implementation of discrete AE signal is similar to the implementation of electronic devices shot noise.

Characteristics of AE signals depend on many factors, primarily on the physicomechanical and acoustic properties of the monitored object, its geometry, the nature of the external load, and electroacoustic receiver characteristics. Basic measured parameters (GOST 27655-88) of AE signals are given in Table 4.1. Many companies and organizations are involved in the development and production of acousto-emission diagnostic systems, among them the largest are Dunegan/Endevco, Trodyne, PAC, DWC (USA); Brüel & Kjær (Denmark); AVT (Great Britain); Setim (France); Vallen-Systeme Gmbh (Germany); JSC "Introscope" , "VNIINK" (Moldova); LLC "NDT In-marriage" (Belarus).

Modern acoustical emission systems are digital. They provide recording of AE signals and measurements of individual parameters of AE signals, and their probabilistic characteristics—the distribution of instantaneous values, correlation and spectral characteristics (Table 4.2).

The analysis of the set of parameters of acoustic emission signals sequence allows determining the location of the source, its type and degree of danger.

Table 4.1 Basic measured parameters of AE signals

Parameter	Symbol	Determination
Number of AE pulses	N_Σ	Number of recorded pulses of a discrete AE during the observation time interval
Total account of AE	N	Number of recorded exceedances by AE pulses of established discrimination level by the observation time interval
AE activity	Σ	Number of registered AE pulses per time unit
AE counting rate	$\dot{N}$	The ratio of the total account of AE to the observation time interval
AE energy	E	The energy released by the source of AE and transported by waves that arise in the material
Classification parameter	n	Indicator of degree in an expression describing the dependence of the total AE count N on the intensity of stress K, $N = aK^n$, where a is a constant reflecting the test conditions

Table 4.2 Comparative characteristics of acoustic emission control and diagnostic systems

Name, manufacturer	Number of channels in block	ADC parameters	Frequency range (kHz)	Dynamic range (dB)	Own noise level (μV)	Number of registered acts per second	Number of registered parameters
AMSY-5, Vallen-Systeme (Germany)	36 (up to 254)	16 bit, 10 MHz	10…2000	82	3	30 000	
DiSP, "Physical Acoustics Corporation/MISTRAS Holdings Company"	8…52	16 bit, 10 MHz	10…2100	82	3	10 000	10
GALS-1, **OKO Association** (Ukraine)	1…100	16 bit, 2.5 MHz	10…800	95	5	52 000	11
AEC-USB, Introscop (Moldova)	1…32	–	10…500	–	5	–	7
Aline-32D, LLC "NTERIUNIS" (Belarus)	64	16 bit, 2 MHz	1…500	84	5	15 000	7
LOKUS-D, ZAT "ELTEST" (Russia)	4…32 (up to 80)	1 MHz	25…200	80	4	–	8

Table 4.2 shows that basic universal module of acoustic emission diagnostic systems should contain up to 8 channels, the full dynamic range should be not less than 60 dB, the lower limit of frequency range lies in the range of 20…50 kHz, the upper limit is 2…3 MHz.

Systems of acoustic contact leakage [3]. Acoustic leakage signals are a consequence of hydrodynamic processes occurring in leakage of pressure pipelines under pressure drop influence. The properties of acoustic leakage signals essentially depend on the fluid flow regimes, which are determined primarily by the magnitude of pressure difference and geometric parameters of the gap.

There are the following basic modes of liquid outflow into the air—drip, continuous filling channel, cavitation, complete flow separation from the duct walls, intracanal jet disintegration.

Drip leakage. For water at a pressure drop of 0.75 MPa and a slit diameter of less than 0.04 mm, there is no movement in the channel and, as a consequence, any acoustic signals.

The movement of a liquid in a leak begins when there is a pressure drop sufficient overcome the surface tension. In this case, either individual droplets of liquid or individual gas bubbles are formed in the outlet section.

Cavitation regime occurs with increasing pressure drop. For example, cavitation occurs in short slots of 5…10 mm with a slit diameter of 0.1…0.1 mm with pressure drops of about 0.1 MPa.

In the vicinity of the initial flow, liquid is separated from the walls of the channel and the formation of cavitation bubbles—caverns. The cavitation flow regime due to instability of the tail part of cavity is accompanied by strong acoustic signals, which exceed the noise of turbulence by an order of magnitude.

With a further increase in the pressure drop, an increase in the length of cavity, its exit from the end of channel and destruction are observed, as a result of which the regime of complete detachment of the liquid jet from the walls begins. At the same time, acoustic signals are minimal.

The mode of intracanal decay of a jet arises only in individual cases and for sufficiently large pressure drops. In this mode, the liquid jet is unstable, its curvature and decomposition into individual drops (spraying) could occur. If the disintegration of jet occurs in the middle of the channel and the trajectory of particles motion of unstable jet or liquid droplets that are detached from the jet cross the channel walls, then significant acoustic signals emerge, the cause of which is the bombardment of the channel walls by separate fluid particles.

Useful acoustic signals (pseudo-sound) could also appear in test object wall in case of small flow velocities due to pressure pulsations on the channel wall caused by flow nonstationary.

For acoustic contact leakage, the cavitation and continuous channel filling modes are of greatest interest. In these modes, the acoustic signals generated by turbulence and cavitation have a broadband continuous spectrum with an upper frequency of several megahertz. In the low-frequency part of the range (up to 60 kHz), the spectrum of acoustic signals, recorded during a leak, can have local maximum corresponding to the natural frequencies of the gap.

Methods and means of acoustic contact leakage are based on the analysis of characteristics and parameters of acoustic leakage signals, which registration is carried out by means of receiving converters that have direct contact with the wall of monitoring object. Basic characteristics of leak detection are accuracy of location and control distance. The accuracy of location is the leak locating error, distance—the possible maximum distance from the leakage point to the receiving converter.

Most of the known devices and systems of acoustic contact leakage are based on acoustic leakage signals correlation processing. The principle of these devices is based on measuring the maximum mutual correlation function time delay between acoustic signals recorded by two separated receiving converters. The distance from the leak to one of the converters is calculated by the formula considering found delay, the distance between sensors and measured or specified sound velocity of the pipeline.

Some devices include acoustic leakage signals processing in frequency domain. Based on spectra and coherence function analysis, an informative area is determined, after which time signal processing is performed in the time domain.

In Ukraine the instruments for acoustic leakage systems are mass produced in MP "DISIT" of the National Academy of Sciences of Ukraine and in the E. Pukhov Institute of Modeling Problems in Power Engineering of the NAS of Ukraine. Overseas leak detectors are produced by Seba KMT (Seba Dynatronic); Intereng Messtechnik Gmbh, FAST (Germany); Primayer Ltd, Palmer Environmental (United Kingdom); Gutermann (Switzerland); Metrovib (France); Fuji Tecom Inc. (Japan) Echologics Engineering (Canada) and others.

Table 4.3 Comparative characteristics of leak detectors

Name, manufacturer	Location accuracy (m)	Frequency range (kHz)	Pipeline diameter (m)	Distance between sensors (m)	Number of sensors
SeCorr-08, InterEng Messtechnik GmbH (Germany)	–	0.001…10	–	–	2
MicroCorr-6 DKL 1506, Seba KMT (Germany)	–	0.005…5	25…1500	250	2
Correlux P-250, Seba KMT (Germany)	–	0…4	–	–	2
LC-2500, Fuji Tecom Inc., (Japan)	–	0.02…5	–	–	2
Eureka3, Primayer Ltd., (Great Britain)	–	0.001…22	–	–	2
KORSHUN—11, "Dysyt" NANU (Ukraine)	± 0.1	0.005…4.5	to 1400	–	2
K-10.3M, Pukhov IMEE NASU (Ukraine)	± 0.5 (0.95 prob.)	–	to 1200	100	3
TKP4102, "Inkotes", (Russia)	0.1% of leakage size	0.001…9	50…1200	–	2

Table 4.3 shows basic characteristics of known leak detectors, which are designed to search the fluid leaks in steel or cast-iron pressure pipelines with a minimum diameter of 20 mm (dash means that information is not available). The required minimum pressure drop between the pipeline and the environment should be at least 0.2 MPa, the minimum for detecting the through opening defect is not less than 0.1 mm.

These leak detection devices are portable and consist of several modules—piezo-electric receiving converters, data transmission channel, measuring station for data collection, information processing unit. Pre-amplifier is built in sensor or made as a separate unit. Dynamic range is 60 … 70 dB, operating temperature: −30 … +70 °C.

Vibrodiagnostic systems [4]. Vibrodiagnostics is one of the main and most promising methods for diagnosing heat and power equipment units, in particular gas turbine and steam turbine installations, power generators, electrical machines, pipelines.

This method is based on the study of characteristics of vibroacoustic signals and their parameters, which are the most sensitive to changes in technical status of objects, the appearance and development of damage, and could be determined directly on operating object.

Vibration parameters are frequency of periodic vibrations, amplitude and period of vibrations, peak value and mean square value of the vibrational quantity.

Primary values characterizing vibration are vibro-displacement, vibration velocity and vibration acceleration.

Modern instruments and vibrodiagnostic systems are based on the probabilistic analysis of vibroacoustic signals, in particular, on the analysis of the waveform and its distribution, spectral, correlation, cepstral analysis, wavelet analysis, etc. The most common is the spectral analysis of vibrations that allows detecting various defects in thermal mechanical equipment elements—imbalance of rotor, impeller; misalignment of shafts; non-rigid fastening; defects in electric motors, compressors, pumps, fans, drive couplings, gear and belt drives, rolling and sliding bearings.

Vibrodiagnostics is carried out with the help of specialized equipment—vibrodiagnostic equipment, the requirements for which are determined by relevant regulatory documents. On the modern vibration diagnostics equipment market, an important role is played by a vibration analyzer—portable instrument for direct measurement of vibration parameters and for vibration signals processing. On the Ukrainian market, the most popular are vibration analyzers produced in Ukraine and Russia: NPP "Contest" , ITC "Vibrodiagnostika" (Ukraine); INCOTES LLC, Interpribor SPE, Vibro-Center PPF, "DIAMEH 2000" (Russia); and also Brüel & Kjær (Denmark); Emerson, Fluke Corporation (USA).

All vibration analyzers, presented in Table 4.4, perform spectral analysis of signals in the frequency range from single Hz to tens of kHz using a fast Fourier transformation with spectrum number from 50 to 51 200. Vibro-analyzers' displays are manufactured using electronic tubes (VGA), liquid crystals (LCD, TFT) or LEDs (LED) technology. Individual devices (VD-1852, AP1013) use a computer monitor. All vibro-analyzers have PC communication interface, most commonly USB.

Thus, diagnostic signals' statistical analysis, in the most modern passive acoustic systems of functional diagnostics, is based on correlation-spectral methods that are comprehensive for Gaussian signals. However, analysis of theoretical and experimental studies results has shown [1] that noise and rhythmic signals, as a rule, are non-Gaussian random processes and often have a uniform spectral density. This limits the capabilities of existing passive diagnostic systems based on methods of correlation and spectral analysis.

One of the promising directions for further development of methods and means for thermal power facilities monitoring is improving existing and developing new passive diagnostic systems based on modern methods of the random processes theory and statistical processing of noise and rhythm signals.

Creation of new passive diagnostic systems for determining the technical status of heat and power equipment elements, increasing their sensitivity and reliability requires the solution of the following main tasks:

Table 4.4 Comparative characteristics of vibration analyzer

Name, manufacturer	Number of channels	Frequency range	Quantity of spectral lines	Screen, resolution	Weight (kg)
2260, Brüel & Kjær (Denmark)	2	8…20 000	>400	PC, 192 × 128	1.2
CSI 2140, Emerson (USA)	4	0…80 000	100…12 800	TFT, 640 × 480	1.79
Fluke 810, Fluke Corporation (USA)	4	2…20 000	800	VGA, 320 × 240	1.9
VD-1852, ITS “Vibrodiagnostika” (Ukraine)	2	0.3…40 000	50…12 800	PC	0.8
795MC911, NPP “KonTest” (Ukraine)	2	2…10 000	400…6400	PC, 160 × 160	1.4
EKOFIZIKA-110A, “OKTAVA—Elektrondizain” Group (Russia)	1	25…48 000	200	OLED, 320 × 240	0.55
Kvartz/Topaz-B, “DIAMEH 2000” Ltd. (Russia)	1	0.3…40 000	100…1600	PC, 240 × 128	2.5
Oniks, “DIAMEH 2000” Ltd. (Russia)	2	0.5…40 000	–	WVGA, 800 × 480	2.5
STD-3300, **“**Technekon” Ltd. (Russia)	2	0…32 000	3200…25 600	LCD, 320 × 240	0.7
SD-21, “Assosiation VAST” (Russia)	2	0.5…256 000	400…1600	LCD, 320 × 240	0.7
ADP-3101, INKOTES Ltd. (Russia)	4	0.5…20 000	200…16 000	VGA, 320 × 240	2
SVAN 958, “Algorithm-Acoustic” (Russia)	4	0.1…20 000	400…1600	LCD, 128 × 64	0.5
ViAna-4, “Vibro-Center” Ltd. (Russia)	4	3…10 000	to 51 200	TFT, 640 × 480	2
Diana-2M, “Vibro-Center” Ltd. (Russia)	2	3…10 000	to 51 200	PC, 320 × 240	1.5
Vibran-3, “Interpribor” (Russia)	4	0.5…1000	0.5…1000	–	0.34

- construction of adequate mathematical models of noise and rhythmic signals accompanying the operation of heat and power equipment elements and reflect the physics of their occurrence;
- determination of the most informative characteristics and parameters that allow monitoring and diagnostics of technical condition of heat power equipment elements;
- development of statistical methods and software for experimental determination of new informative characteristics and parameters.

4.2 Diagnostic System for Electric Power Facilities

A complex technical object can almost always be viewed in the form of a certain hierarchical structure [5]. This is especially evident in the case of the modern electric power system, where different organizational levels "from top to down" could be distinguished: from the central dispatching control and to construction elements of power generating equipment or electric networks, or in household electrical appliances of end user.

From the point of view of developing systems for diagnosing and monitoring the condition of electric power equipment, it is necessary to confine ourselves to the structure of a particular power plant, distribution substation, power line, etc. [6].

As an example, Fig. 4.1 shows the hierarchy of electrical equipment of a traditional power plant (TPP, NPP, HPP) that is considered from the point of view of technical diagnostics.

At the first, the lowest level, construction elements of the power plant equipment main units are located. It is this level that determines what defects are possible in

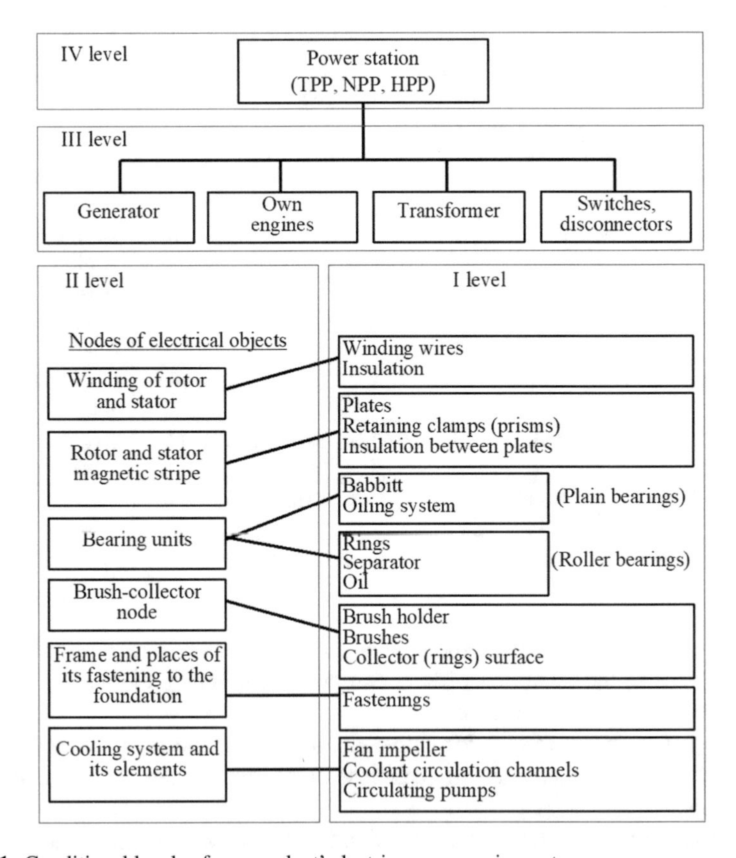

Fig. 4.1 Conditional levels of power plant' electric power equipment

studied object. Deep study of the elements located on the first level of hierarchy gives all necessary information about types, causes and appearance of defects. As a result of this analysis, diagnostic models are constructed, diagnostic signals and parameters are selected.

The second level is equipment nodes that are constructively integrated whole. This includes rotor winding and stator rotating machines, magnetic circuits, bearing units, housing, bed, foundation, cooling system.

The third level represents electrotechnical equipment of power plants, generators, engines of own needs, transformers, switches, disconnectors, insulators, pumps, etc.

The fourth level of hierarchy is the level of power plant as a whole.

Higher levels could also be considered: energy integration, the country's energy system etc.

The structure of the diagnostic system being developed can be conditionally divided into hierarchical levels, similar to the way it was done above with regard to the power plant equipment (Fig. 4.2).

The distribution of functions between these hierarchical levels should be organized as follows:

- level I—primary selection of diagnostic information (measurement of diagnostic signals, amplification, analog filtering, digitization);
- level II—primary mathematical processing and the adoption of intermediate diagnostic solutions (simple algorithms, which implementation does not require significant computing resources, information separation by the degree of defects criticality), signaling to a higher level in the presence of appropriate prompt;
- level III—collecting, complete processing deep data analysis, fast response to lower level alarm, adoption of diagnostic solution about diagnosed object as a whole, statistical data archivation, reliability prognosing and equipment residual life estimation, repair works planning.
- level IV—presenting data to different users (including geographically remote ones, for example via Web technologies) with access rights restriction depending on official duties.

It is advisable to combine the functions of levels I–III into a separate subsystem for each large object that is part of the power plant, for example—for each power unit, each powerful circulating pump, each transformer, etc., that is, for a certain set of structurally and logically combined units, forming a single whole and fairly compactly located in space. Therefore it is logical to call the system of this level a local technical diagnostic system (TDS).

The higher level IV, in turn, combines information flows coming from different level III systems. Therefore, it is advisable to call the level IV "central".

According to the structural scheme above, at the top level of hierarchy of an intelligent distributed multi-level system for monitoring the status and diagnostics of electrical equipment is the central module—central diagnostic system (CDS). The CDS can be used to plan the operation of an object as a whole, design different kinds of reports etc. It is assumed that operating conditions of the CDS equipment are office premises, that is, there are no significant levels of harmful external influences,

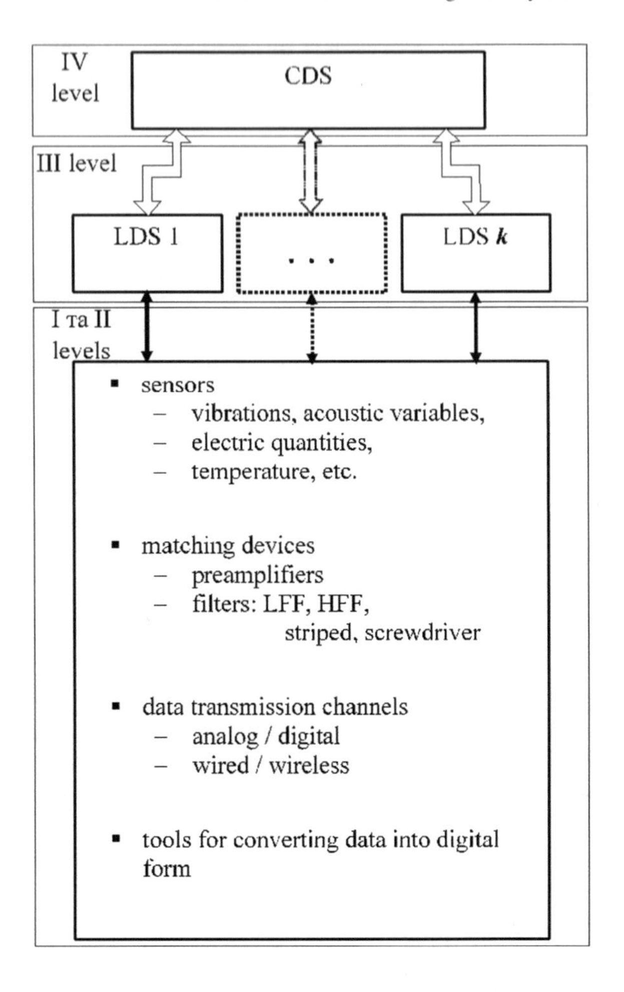

Fig. 4.2 Hierarchical organization of TDS for power plant

the operating temperature range corresponds to usual operating conditions, electric power is possible to take from the public network.

Low levels of system hierarchy are represented by components of the local diagnostic system (LDS). The equipment, which is part of the LDS, must operate under conditions of electric power facility production process. Such equipment must be protected from the effects of significant levels of electromagnetic interference, allow operation over a wide range of temperatures, and, possibly, in corrosive environments.

First of all, this requirement concerns measuring transducers, but it can also be extended to the whole list of equipment. This equipment belongs to I and II levels of hierarchy according to the classification adopted above.

Separately, we should consider the requirements for PC corresponding to the III level of hierarchy. Depending on the specific production conditions, this unit can be located both in production zones (then special requirements apply to it) and in station operative personnel room (in this case there are no special requirements).

The functions, the LDS must perform, require this system to be built on the basis of digital computational tools. This applies to equipment of all hierarchical levels, except for measuring transducers, since most of the physical processes that need to be measured have an analogous nature.

Considering various factors (quite significant computing power of the CPU with "overclocking" possibility, significant RAM size, the availability of a large number of digital interfaces, the possibility of using a removable drive in the form of microSD memory card, the possibility of using widespread free Linux operating system, low power consumption, small overall dimensions, low price and high manufacturing quality), as a basic computational component for modules of II hierarchical level of the intelligent distributed system for diagnostics and monitoring the electric power equipment was chosen as a single-board computer of the Raspberry Pi Model B+ type. Basic technical parameters of this computer are given in Table 4.5 (data taken from the official site of the developer Raspberry Pi Foundation company, http://www.raspberrypi.org/).

There are a number of technological solutions on the market that have already become the standards of wireless digital communication and are used to exchange digital information between objects at a distance. Series of standards IEEE 802.11 (WLAN, Wi-Fi), the standard 802.15.1/a (Bluetooth), and the standard 802.15.4 (ZigBee) are widely applied [7]. All these standards are designed to use the 2.4 GHz

Table 4.5 Basic technical characteristics of a single-board computer Raspberry Pi Model B+

Parameter	Value
Model name	Raspberry Pi Model B+
Producer	Raspberry Pi Foundation, GB
Start of sale	July 2014
Hardware platform	Broadcom BCM2835
CPU	700 MHz, ARM1176JZF-S, ARM11 ARM architecture v6
RAM	512 МБ
Drive	Slot microSD, до 64 GB
Operating systems	Linux (Raspbian, Debian, Fedora, Arch Linux)
USB	4
Number of output leads	40
Interfaces	26 × GPIO, UART, I^2C bus, SPI bus
Network interface	Ethernet, 10/100 Mbps
Power supply	5 V, microUSB
Power consumption	650 mA, ≈3 W
Dimensions	85 × 56 × 17 mm

Table 4.6 Comparison of basic wireless standards parameters

Standard	802.15.4 ZigBee	802.15.1 BlueTooth	802.15.3 HighRate WPAN	802.11b Wi-Fi
Direction of application	Monitoring, control of sensor network, home/industrial automation	Voice, data, cable replacement	Streaming multimedia, replacing audio/video cable	Data, video, local area networks
Main advantages	Price, energy saving, network size, less loaded ranges	Price, energy saving, frequency hopping	High speed, energy saving	Speed, flexibility
Carrying frequency	2.4 GHz			
Maximum data transfer speed	250 kBit/s	1 MBit/s	22 MBit/s	11 MBit/s
Capacity (dBm)	0–10	0 (class 3) 4 (class 2) 20 (class 1)	0	20
Coverage radius (m)	100–1200	10 (class 3) 100 (class 1)	5–10	100
Sensitivity (dBm)	−85	−70	−75	−76
Network size (Kb)	4–32	>250	–	>1000

radio frequency band, and do not require licensing for use around the world [8, 9]. Table 4.6 gives, for comparison, basic technical indicators of these standards.

Considering significantly longer range of stable communication compared to rivals, the maximum number of network elements, the ability to self-organize and self-repair of network, an advantage in building intelligent distributed diagnostic systems should be given to ZigBee standard. Devices implemented on the basis of the ZigBee standard have a power saving mode (sleep mode) that provides significantly longer (up to 2 years) battery life of such devices compared to batteries of another rivals. ZigBee device can respond to an external request from sleep mode in less than 30 ms (compared to 3 s required for Bluetooth-based devices). This factor is essential in the construction of rapid response facilities, which, obviously, should be part of the diagnostic system.

Table 4.7 provides a list of some ZigBee wireless digital communication modules with USB interface. From their technical parameters comparison, it can be concluded that in an intelligent distributed diagnostic system a module of EMB-250-100BI-U-007 type from EMBee (Ukraine) and a module of ZE51-2.4 RF USB type from Telit Wireless Solutions (USA) are suitable for application. Module of IA OEM-DAUB1 2400 type by Integration Associates (USA) has a very short communication range,

Table 4.7 ZigBee wireless communication modules

Manufacturer, model, photo	Basic technical parameters specified by the manufacturer
EMBee, Ukraine EMB-250-100BI-U-007	Frequency range 2.4 – 2.48 GHz Communication range: to 2000 m (direct visibility) to 100 m (indoor) Max output power 100 mW Price – to order
Telit Wireless Solutions, USA ZE51-2.4 RF USB Dongle	Frequency range 2.4 – 2.48 GHz Communication range: to 1000 m (direct visibility + antenna) Max output power 2,5 mW Data speed: 115.2 kbps Price – £47,40 (without taxes)
Integration Associates, USA IA OEM-DAUB1 2400	Frequency range 2.405 – 2.48 GHz Communication range: to 30 m (direct visibility) Max output power 10 mW USB 1.1 Interface Price – £28,00 (without taxes)

comparing to much cheaper modules of Bluetooth standard, and can't be used in this case.

Figures 4.3 and 4.4 show laboratory samples of modules that are a part of the distributed hierarchical system of wind power unit (WPU) diagnostics, and Fig. 4.5—location of the data acquisition unit from WPU low-shaft rotating parts.

According to the developed generalized structure of an intelligent distributed multi-level system for monitoring the status and diagnostics of electric power facilities, a prototype system that contains a number of units on various levels, built on single-board miniature computing units and personal computers, was produced.

Algorithmic software managing interoperation of modules of the system was developed. Experimental check in demonstration mode showed operability of the developed prototype system, in particular, the possibility of reliable two-way data exchange between individual system modules.

Fig. 4.3 Portable computer and ZigBee module

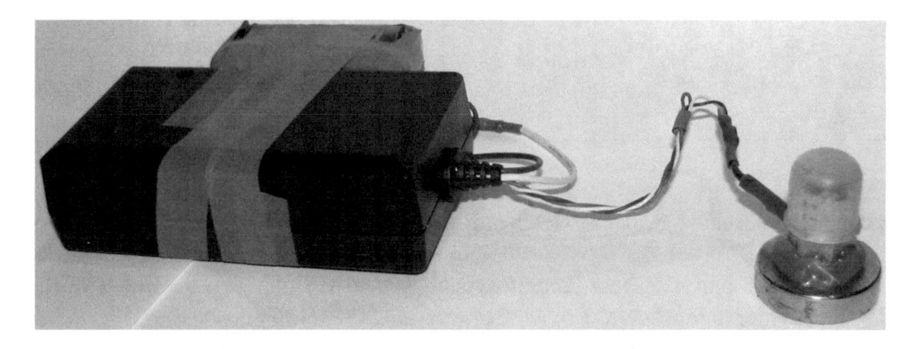

Fig. 4.4 Module with autonomous power for recording vibrations on moving parts

4.3 Diagnostic System for Heat Power Facilities

Maintenance of operational reliability, durability and safety of heat and power equipment is a complex task connected with the organization of reliable control of power plants and ensuring optimal conditions for their operation. To solve this problem, it is necessary to have special monitoring systems that allow monitoring of the heat engineering processes of generation, transportation and consumption of heat energy;

Fig. 4.5 The registrator-transmitter (module of data collection from rotating parts), mounted on WPU shaft

measuring basic parameters of heat-power installations, equipment, machines, mechanisms, etc.; diagnosing and predicting the technical condition of installations and their units [10–13].

Basic parameters of the heat and power equipment, which is diagnosed, include:

- general parameters—economic coefficients related to technological process factors;
- metal structures characteristics—hardness, creep, fracture resistance, shells presence, impurities, scale formation of heating surfaces;
- structures geometrical parameters—diameter and thickness of pipes, relative displacements of individual units;
- thermophysical processes parameters—temperature of overheating zones of heating surfaces and vapor lines;
- chemical processes parameters—condition of cooling media water;
- noise processes parameters—appearance of acoustic emission signals, acoustic leakage signals, boiling fluid noise, noise in pipelines, etc.;
- vibration parameters—vibration of boiler, pipelines, fans, smoke exhausters.

To solve the problems of big thermal power systems monitoring and diagnosing, it is expedient to use the system approach methodology. One of its main provisions is the separation of several levels of hierarchy in the heat energy system. Figure 4.6 shows the hierarchical structure of the thermal power system of a large industrial

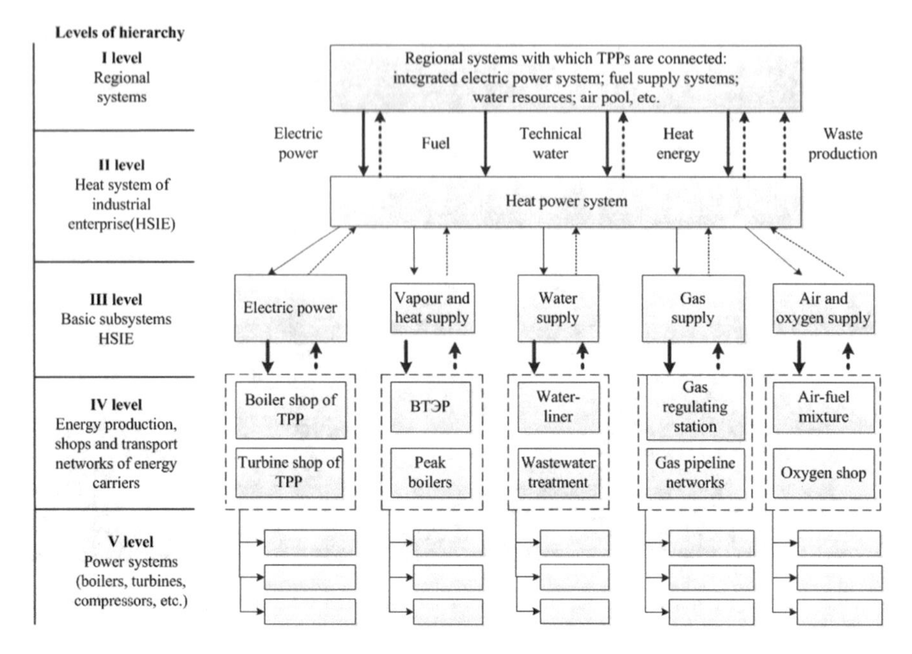

Fig. 4.6 Hierarchical structure of the thermal power system of a large enterprise

enterprise. Elements of the V-level are complicated themselves (for example, a vapor turbine) and could be subject to further detalization at lower levels.

The tasks of hierarchical levels II–IV include, for example, distribution of different types of fuel between individual consumers; selection of composition and profile of basic power equipment; optimization of parameters and type of thermal scheme of HSIE, etc. The tasks of V-level and lower hierarchical levels include the selection of optimal thermodynamic and design parameters of a specific heat power equipment with the parameters specified at levels II–IV [14].

This approach to the consideration of the heat and power system allows using of Smart Grid technology for individual levels diagnostics.

The emergence and development of the Smart Grid concept is a natural stage in the heat and power system evolution, caused on the one hand by the obvious needs and problems of modern heat and power market, and on the other hand by technological progress, primarily in the field of computer and information technologies.

The existing thermal power system without Smart Grid could be characterized as passive and centralized, especially the last circuit part—from distribution networks to consumers. In this part of the heat supply chain Smart Grid technology significantly changes the operating principles, offering new approaches to active and decentralized interaction.

Smart Grid technology (Fig. 4.7) is characterized by several innovative properties, corresponding to new market needs, among them:

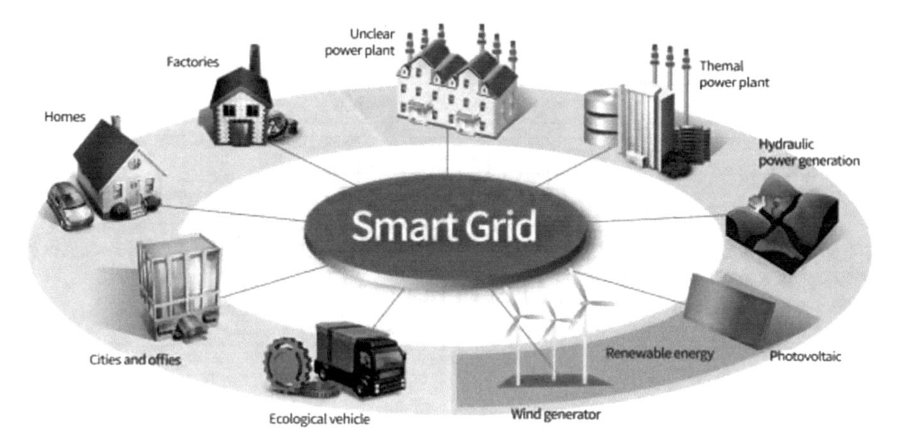

Fig. 4.7 Schematic diagram of Smart Grid technology application directions

1. Active bi-directional scheme of interaction and information exchange in real time between all elements (network participants)—from thermal energy producers to end users.
2. Coverage of the whole technological chain of the system: producers of thermal energy, distribution networks and end users.
3. To ensure data exchange Smart Grid uses digital communication networks and data exchange interfaces. One of the most important goals of Smart Grid is to provide an almost continuous managed balance between demand and supply of thermal energy. To do this, network elements must constantly exchange information about the parameters of thermal energy, consumption and generation modes, the amount of energy consumed and planned consumption, and commercial information.
4. Smart Grid is able to effectively defend itself and self-recover from major disruptions, natural disasters, external threats.
5. Smart Grid helps optimal operation of heat and power system infrastructure.
6. From the overall economy point of view, Smart Grid promotes the emergence of new markets, players and services.

Smart Grid technology works through a system of special "smart" sensors installed in enterprises and in residential areas. They inform about the level of thermal energy consumption, which allows adjusting the use of heat engineering equipment in time and distribute heat energy, depending on the needs.

The newest information systems in the energy sector cover big data sets, Internet, wireless data transmission networks, "cloud" computer technologies, etc. In particular, the use of wireless sensors and other hardware equipment for thermal power plants has increased significantly in recent years. They measure and transmit to the control desk a large amount of information: temperature, pressure in the pipeline system, vibration characteristics and others. Having this information from wireless sensors, which in many cases can not be obtained in any other way, monitoring and

Table 4.8 Comparison of traditional and new approaches to the maintenance of thermal power equipment and ensuring its reliability

Traditional approach	Smart Grid
• Functional diagnostics (continuously or periodically) only for particularly sensitive objects • System of planned and preventive repairs • Diagnostic test (during scheduled stops) • Local systems for diagnostics, protection and automation for particularly important facilities	• Condition diagnostics and remote monitoring for a wide class of equipment • Maintenance and repair by actual condition • Adaptive distributed security systems (diagnosis, monitoring status, self-recovery)

diagnostic systems over a certain period can more effectively assess the need for preventive maintenance work on heat and power equipment [15].

Proceeding from the foregoing, the concept of Smart Grid essentially changes the requirements to reliability of heat and power networks equipment, and accordingly also requirements to means of its maintenance (Table 4.8).

In particular, within the traditional approach, equipment maintenance was carried out on the basis of preventive maintenance, and technical diagnostic tools were used to find defects after the object was taken out of work. Particularly important tools are equipped with their own controlling and monitoring systems that provide emergency signaling in the event of contingencies, but have insufficient means to identify, classify and localize defects.

Within the Smart Grid concept framework, it is assumed that maintenance and repairs will be carried out according to the actual status. To do this, largest part of the equipment will be covered by reliability assurance systems that will perform constant or periodic monitoring of its actual technical condition. In addition, these systems will have more possibilities: two-way information exchange at all levels, remote condition monitoring, failure prediction, spare parts planning, residual resource assessment, self-recovery of equipment [16].

In foreign literature, the above tasks are united under the general title of "Asset Management". Now both engineering and scientific works are being actively conducted in this direction, and their authors connect their results with the realization of the Smart Grid concept key moments. Leading manufacturers of heat and power equipment already now offer a number of software products designed to collect and summarize statistical information about operating conditions and the actual status of heat and power networks various equipment.

The need to equip a wide class of diverse heat and power equipment with diagnostics, monitoring and condition control systems means that these systems must be adaptive, more intelligent than existing ones. An important role in ensuring the broad capabilities of next-generation systems will be the distribution of computing resources between various diagnostic, monitoring and control systems operating at different levels of heat and power system hierarchy.

The essence of the developed system for diagnosing heat and power equipment is to monitor and make diagnostic decisions at each of the individual hierarchical levels, which allow identifying, localizing and eliminating defects before the diagnostic objects become faulty.

Based on HSIE equipment hierarchy, the system measures diagnostic signals that carry information about the actual status of equipment units that are diagnosed. Thus, the system can include sensors of those physical quantities that are used to diagnose a specific system. Depending on the object of diagnosis, the system may include [17]:

- thermocouples or thermistors—for measuring temperature;
- accelerometers—for measuring vibration parameters;
- measuring microphones—for acoustic noise level determination;
- electrical quantities sensors—for measuring the parameters of transformers functioning;
- pressure sensors—for monitoring the depression in the furnace;
- gas sensors—for determining the concentration of harmful substances in the smoke path;
- thermal energy meters—for determining the current operating mode of heat engineering equipment, etc.

Modern diagnostic systems are almost always built on the basis of some digital calculation means (microcontroller, personal computer, industrial workstation, etc.). For a diagnostic system that conforms to the basic principles of Smart Grid concept, this requirement is mandatory, since in the framework of smart networks, information is exchanged in digital form. Thus, measured signals must be digitized for further processing in the computing core of the system. The final stage of information processing within the diagnostic system is the reflection of the results to users. For this the system structure includes appropriate tools, which, in particular, should provide authorization for system users, access rights distribution, and information protection.

It should be noted that a significant number of diagnostic signals could be measured in complex diagnostic systems, which leads to a huge exchange of information between system components. To reduce the load on communication channels, decentralization principle of computing resources is applied, which is one of the basic Smart Grid concept principles.

Thus, the structure of the diagnostic system that is being developed can be conditionally divided into hierarchical levels, similar to the way it was done above with heat engineering equipment of the heat and power system (Fig. 4.8).

The distribution of functions between hierarchical levels of the system under development is expediently organized as follows:

- level I (Measuring Transducers (MT))—primary selection of diagnostic information (measurement of diagnostic signals, amplification, analog filtration, digital conversion);
- level II (LDS)—collecting, complete processing deep data analysis, fast response to lower level alarm, adoption of diagnostic solution about diagnosed object as a whole, statistical data archivation, reliability prognosing and equipment residual life estimation, repair works planning;

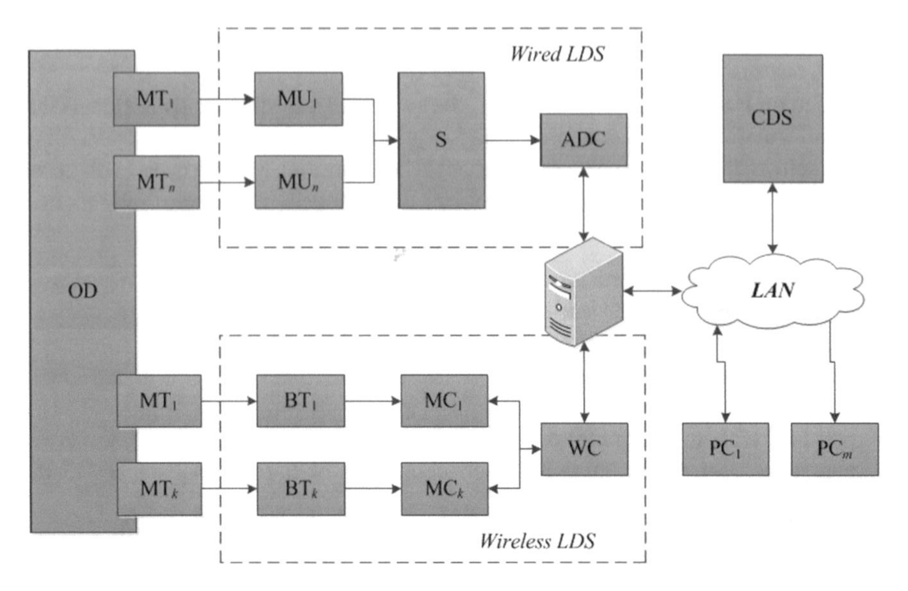

Fig. 4.8 Structure of the multi-level system of heat engineering equipment diagnostics

- level III (CDS)—data representation for various users (including geographically remote, for example, through Web technologies) with restricted access rights depending on their official duties.

To display information for local users (for example, maintenance personnel), as well as to exchange information with the central TPS diagnostic system, all LDSs are included in an Ethernet-based LAN.

To enable information exchange with external users (this could be both people and devices operating outside of this TPS, but integrated into a "smart network"), the CDS has a connection to the global network (Internet). Thus, a number of serious problems arise in ensuring information security and preventing possible terrorist attacks. To solve these problems, special network security hardware is used.

The system for heat engineering equipment diagnostics can work both with wired and wireless LDSs. Wired LDS consists of a matching unit (MU), a switch (S), an analog-to-digital converter (ADC) and PC. Wireless LDS consists of a block of transformation (BT), a microcontroller (MC), a wireless communication (WC) and PC. The use of both wired and wireless LDSs can significantly expand the classes of heat and power equipment that is diagnosed.

Consideration of the degree of critical defects at the stage of system development makes it possible to simplify its structure; reduce the amount of information that is processed in the system and transmitted between its hierarchical levels; and ultimately reduce system cost while maintaining its functionality at an adequate level.

The main advantages of the proposed system for diagnosing heat energy equipment based on Smart Grid are:

- reliability (Smart Grid prevents massive heat off);
- security (Smart Grid constantly monitors all elements of the network in terms of their operation safety);
- energy efficiency (reduction of thermal energy consumption, optimal consumption leads to a decrease in the requirements for generating capacities);
- ecological compatibility (achieved by reducing the amount and power of generating elements of the network, leading to a decrease in the concentration of harmful substances in the surrounding space (CO, NO_x, C_xH_y, H_2, C, etc.).
- financial profitability (operating costs reduction; consumers have accurate cost information and can optimize their heat energy costs; in turn, business can optimally plan operation costs and development of generation and distribution networks).

4.4 Combustion Process Features

Fuel combustion monitoring is reduced to monitoring of waste gases content, while the studied objects are boiler and air-fuel path [18]. The structural diagram of fuel combustion monitoring is shown in Fig. 4.9.

The efficiency of boiler unit is determined by the efficiency of its components: burners, heating surfaces, heat exchangers (economizers, air heaters), draft machines and other devices. One of the most important components of the combustion process is fuel combustion efficiency, that is, the economy of the operation of the burners themselves and associated equipment (fans and smoke exhausters).

The equation of boiler heat balance in general form in stationary mode has the following form:

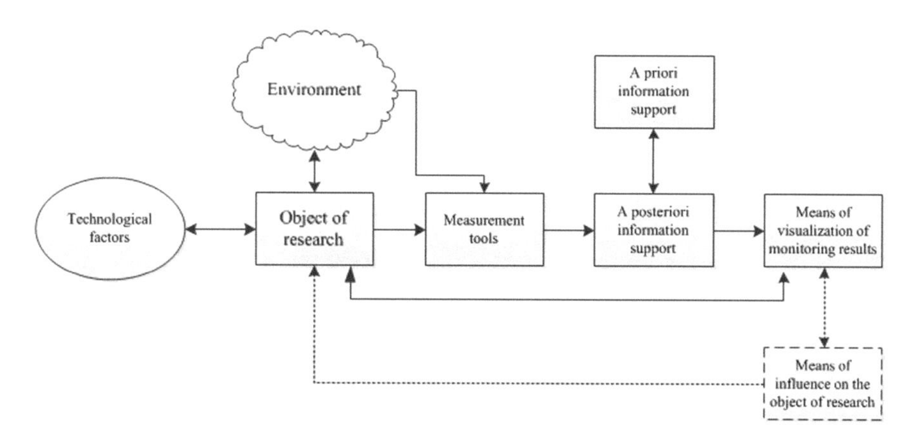

Fig. 4.9 Structural scheme of combustion process monitoring

$$Q_N = Q_1 + Q_2 + Q_3 + Q_4 + Q_5 + Q_6,$$

where Q_N—available heat, Q_1—useful heat, Q_2—heat loss with the exhaust gases, Q_3—heat loss with combustion chemical incompleteness, Q_4—heat loss with combustion mechanical incompleteness, Q_5—heat loss from the heating surfaces, Q_6—the loss from the slag physical heat.

Fuel combustion economy is characterized by the efficiency value that represents the difference between the thermal energy that was released during the fuel combustion and the energy losses in boiler unit. Efficiency can be determined by direct and reverse balance:

- $Q_1/Q_N = q_1 = \eta_d$—efficiency in direct balance;
- $\eta_r = 100 - (q_2 + q_3 + q_4 + q_5 + q_6)$—efficiency in reverse balance.

The main losses of heat during the natural gas combustion are heat losses with exhaust gases; thermal losses associated with combustion chemical incompleteness; heat losses from heating surfaces. The heat losses with exhaust gases depend on: the temperature difference between the exhaust gases and the air supplied to the boiler furnace and the residual oxygen content in the off-gas that characterized the air excess ratio (AER, α) or the air-fuel ratio. These losses are significant (for small and medium-sized boilers they can be from 10 to 26%, for gas boilers and power plants boilers—6–12%) and mainly affect the boiler efficiency.

Figure 4.10 shows heat losses with exhaust gases, calculated by M. B. Ravich method, for different values of exhaust gases temperature.

According to the methodology based on the generalized characteristics of the fuel, during the combustion of natural and associated gases q_2 is determined by the

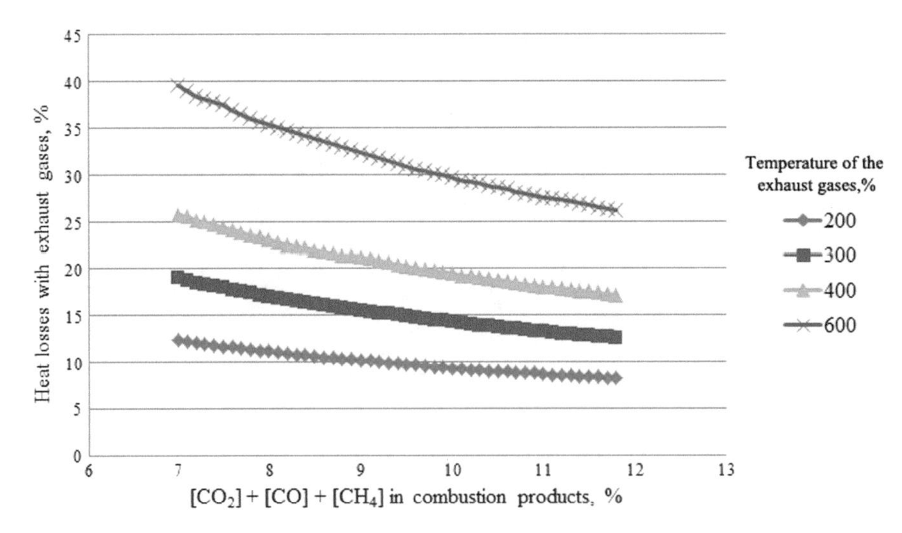

Fig. 4.10 Heat losses with exhaust gases in boiler with different composition of combustion products

formula (%):

$$q_2 = 0.01 \cdot z \cdot (t_g - t_a),$$

where z takes the tabulated value [19], tg—the temperature of exhaust gases, ta—the ambient temperature. At the same time, increasing tg by 10 °C above the normal value for a given boiler load cause increasing q_2 by at least 0.5%, and increasing α by 0.1 cause increasing q_2 by about 1%.

Heat losses with chemical fuel underburning depend on: air excess ratio, the quality of fuel and air mixing; the completeness of fuel combustion and the content of combustible residues in the off-gas ([CO] + [H_2] + [CH]). These losses should be minimized with proper organization of the combustion process.

According to the procedure given above, heat losses with combustion chemical incompleteness q_3 are determined from the data of fuel combustion products according to the formulas:

- $q_3 = \frac{35 \cdot [CO] + 30 \cdot [H_2] + 100 \cdot [CH_4]}{CO_2 + CO + CH_4}$—for natural gas;
- $q_3 = \frac{40 \cdot [CO] + 30 \cdot [H_2] + 110 \cdot [CH_4]}{CO_2 + CO + CH_4}$—for petroleum gas.

Heat losses from incomplete fuel combustion can be significant, and reach the values in the range from 3.5 to 7% (depending on the air excess ratio). In this case, with certain structural features, combustible gases can be burned without loss of q_3.

Heat losses to the environment q_5 include the heat, which is given off by lining and other boiler parts to the environment. The value of q_5 depends on the quality of the lining and insulation of the external walls of the unit and on the temperature difference between its external surface and the environment. For hot-water boilers of the type KV-GM, KVG, TVG, KSVT, KSV, Turbomat, the dependence of heat losses to the environment q_5 from the boiler thermal power is shown in Fig. 4.11.

For different types of boilers the value of the parameter q_5 according to Fig. 4.11 does not correspond to reality in connection with their structural features. Therefore,

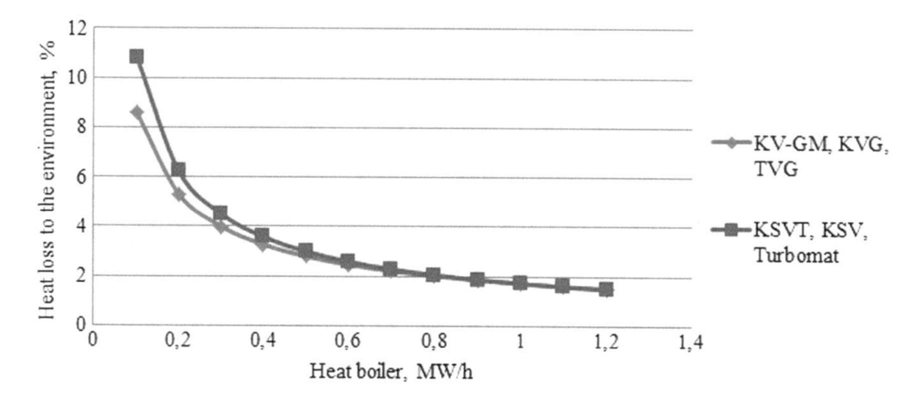

Fig. 4.11 Dependence of heat loss in the environment from the boiler heat output

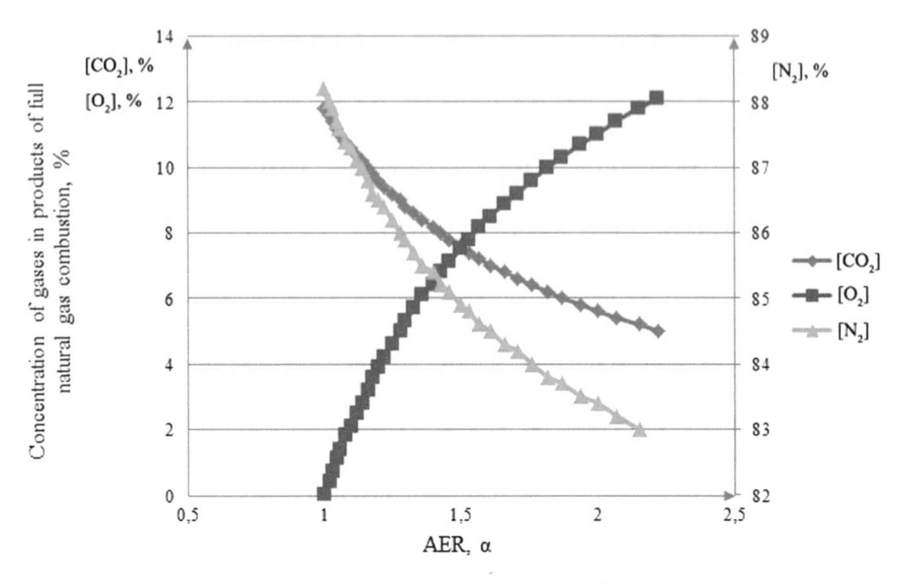

Fig. 4.12 Dependence of the combustion products composition from AER

to compare the functioning of boilers and other thermal units operating with natural gas, there is the use of fuel utilization factor (FUF):

$$\eta_u = 100 - (q_2 + q_3).$$

Figure 4.12 shows a theoretical graph of the dependence of natural gas combustion products from the air excess ratio with complete fuel combustion.

Table 4.9 shows the composition and quantity of methane combustion products (as basic component of natural gas).

The decrease of air excess ratio contributes to: decreasing the oxygen concentration in exhaust gases, increasing the efficiency and, as a consequence, decreasing the temperature of exhaust gases and the electricity consumption by the fan and exhauster.

Simultaneously, the emissions level of harmful nitrogen oxides (NOx) is reduced, which leads to a reduction of environmental pollution. The appearance of chemical underburning (CO) in the fuel combustion products determines the limit of the permissible effect on the air supply reduction. This boundary is unstable and depends on the characteristics of burners and boiler load. Its position is also affected by fuel composition; climatic conditions; temperature of fuel and air; technical condition of the equipment and many other factors. The area of economically advantageous fuel combustion regime is responsible for a low oxygen content (0.5–1.5%) and the appearance of chemical underburning at a level of more than 200 ppm [20].

The optimum composition of the waste gases of the boiler plant is given in Table 4.10.

Table 4.9 Composition and amount of methane combustion products

Indicator		AER, α						
		1	1.1	1.2	1.3	1.4	1.5	2
Air consumption, m^3/m^3		9.52	10.47	11.42	12.38	13.33	14.28	19.04
Quantity of products combustion m^3/m^3								
wet		10.52	11.47	12.42	13.38	14.33	15.28	20.04
dry		8.52	9.47	10.42	11.38	12.33	13.28	18.04
Composition of combustion products, vol. %								
wet	[H_2O]	19.11	17.43	16.10	14.95	13.96	13.09	9.98
	[CO_2]	9.51	8.72	8.05	7.48	6.98	6.54	4.99
	[O_2]	–	1.74	3.22	4.49	5.58	6.54	9.98
	[N_2]	71.38	72.11	72.63	73.08	73.48	73.83	75.05
dry	[CO_2]	11.74	10.56	9.59	8.79	8.11	7.53	5.54
	[O_2]	–	2.11	3.84	5.27	6.49	7.53	11.09
	[N_2]	88.26	87.33	86.57	85.94	85.40	84.94	83.37

Table 4.10 Optimal composition of dry natural gas combustion products

Components	Optimal concentration %	Note
[O_2]	0.5–5	Oxygen
[CO_2]	12–16	Carbon dioxide
[CO]	<0.01	Carbon monoxide
[NO_x]	<0.02	Nitrogen oxides
[CH_4]	<0.01	Methane
[H_2O]	~0	Water in vapour form
Other components	~0	Solid residues

Monitoring of the fuel combustion is carried out on the basis of physical and chemical methods of control, the comparison of which with the help of a universal qualitative efficiency criterion (UQEC) is given in Table 4.11.

Today most heating boilers operate with regime maps that are updated every three years. In these maps, the amount of air supplied to the combustion does not depend on the change in fuel characteristics and equipment condition. During the preparation of the regime map, the commissioners deliberately increase air consumption for combustion, to avoid chemical underburning caused by the lack of stationary instruments for monitoring the waste gases composition, and by the fact that boilers often work with manual regulation of fuel and air supply.

Table 4.11 Comparison of methods for controlling the composition of waste gases based on UQEC

Method	Parameter					UQEC
	Speed	Reliability	Multicomponent	Selectivity	Cost	
Magnetic	0	1	1	0	0	0.4
Thermoconductivity	0	1	1	0	1	0.6
Thermochemical	0	1	1	0	0	0.4
Pneumoacoustic	0	0	0	0	1	0.2
Pneumatic	0	1	1	0	1	0.6
Infrared	1	1	1	1	0	0.8
Hemiluminescent	1	0	1	1	1	0.8
Semiconductor	0	1	1	0	0	0.4
Polarographic	1	1	1	1	1	1
Fluorescent	1	0	1	1	1	0.8
Photolarometric	1	0	1	0	0	0.4
Ionization	1	0	1	0	0	0.4
Thermomagnetic	1	1	1	1	0	0.8
Standard	1	1	1	1	1	1

In addition, the lack of furnace and chimneys hermeticity control leads to increasing of exhauster efficiency through the air inflow from the boiler room. During operation, the operator visually determines the combustion quality, as a result of which the air flow can increase even more and the operating point will shift to the region of large α. All this leads to fuel over-consumption and increased pollutants emission into the atmosphere.

Approach boiler operation closer to the most rational operation modes could be done with gas analyzers (Table 4.12) or combustion process automatic control systems (Table 4.13). Most of the latter is based on the use of a combined informing method about the content of residual oxygen and incomplete combustion products with a cross restriction, which allows more accurate (in comparison with parallel control) maintaining the air-fuel ratio [21].

Such systems reduce the oxygen content in exhaust gases until carbon monoxide appears in them (the optimum is between 50 and 200 ppm) [20]. The appearance of CO in exhaust gases indicates the formation of local zones in the boiler furnace with a chemical underburning of the fuel.

Basic disadvantages of such systems are the following: presence of sampling and sample preparation systems that significantly increases the measurement time; the absence of fans and smoke exhausters frequency regulation that makes it difficult to maintain the optimal fuel combustion mode; long system installation time; relatively long payback period; designed exclusively for large-capacity boilers.

A fundamentally new approach to the fuel combustion monitoring is based on the use of a broadband oxygen sensor (Fig. 4.13) [22].

Table 4.12 Portable gas analyzers

Name	Substances	Setting time, s	Producing country	Cost, $	Foto
PEM-4M2	O_2, CO, CO_2, NO, NO_2, SO_2	120	Russian Federation	on request	
PGA-600	O_2, CO, CO_2, NO_2, SO_2, H_2, CH_4, C_3H_8, H_2S, NH_3, Cl_2	60	Russian Federation	~900	
TESTO-350	O_2, CO, NO, NO_2, SO_2	60	Germany	~4700	
OKSI-5M	O_2, CO, CO_2, NO, NO_2, SO_2	30	Ukraine	~1500	
Green-Line 8000	O_2, CO, CO_2, NO, NO_2, SO_2, C_xH_y	180	Italy	on request	

At present, oxygen probes are widespread in the automotive industry due to the constantly growing strict regulations for the toxicity of exhaust gases. An essential advantage of such probes is the CO oxidation on the surface of the sensor containing ZrO_2. This makes it possible to obtain information on the actual oxygen concentration in combustion products. The disadvantage of their application is the impossibility of detecting chemical underburning in zone of $\alpha > 1$, however, as experimental data showed, supporting the boiler operation with an air excess ratio $\alpha \geq 1.1$–1.15 excludes the possibility of CO formation at a level of more than 200 ppm.

The probe construct assumes the presence of two chambers (cells): measuring and pumping (Fig. 4.13b). Through the hole in pump cell wall, exhaust gases enter the measuring chamber (diffusion gap) in the Nernst cell. This configuration is characterized by a constant maintenance of the stoichiometric air-fuel ratio in diffusion chamber. Supply voltage modulating electronic circuit maintains composition of the mixture corresponding to $\alpha = 1$ in measuring chamber. For this purpose, the pump cell removes oxygen from the diffusion gap into the external medium with a lean mixture and an excess of oxygen in exhaust gases, and, with the enriched mixture and insufficient oxygen, pumps the oxygen molecules from the surrounding medium into the diffusion gap. Current direction during oxygen pumping also differs (Table 4.14).

Table 4.13 Fuel combustion control systems of boiler units

Name	System composition	Disadvantages	Foto
EKO-3	– converter of fan and smoke exhauster drive frequency; – gas pressure sensor on the burner; – CO and O_2 sensors; – throttle sensor in the chimney; – LogicCon control panel; – GPRS modem.	– Internet connection; – technological complexity of development; – significant operating costs during system connection; – designed exclusively for boiler units of high capacity.	
FAKEL-2	– CO and O_2 sensors; – microprocessor regulator "Miniterm-400"; – recorder "Technogra-160"; – vapour flow meter.	– sampling and preparation systems; – absence of fan and smoke exhaust frequency regulation; – relatively long payback period (up to 2 years).	
IT16RN-1	– control system for combustion process SKPG-1m; – block of test preparation GMS-1; – display board di16; – communication and power supply unit CSB-1.	– sampling and preparation systems; – relatively large time of system installation; – relatively long payback period.	
ANGOR	– gas analyzer "ANGOR-C" (O_2, CO); – controller "SPECON SK-2"; – gas pressure sensor on the burner.	– absence of fan and smoke exhaust frequency regulation; – designed exclusively for boiler units of high capacity.	

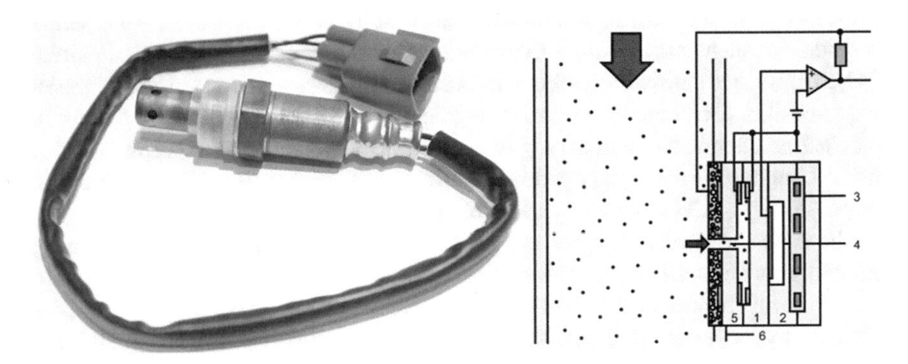

Fig. 4.13 Broadband oxygen probe: **a** appearance; **b** structural scheme (1—Nernst cell, 2—reference cell, 3—heater, 4—diffusion slit, 5—pumping cell, 6—air and fuel tract)

Table 4.14 Amperage dependence on air excess ratio in an oxygen probe

I, mA	−3	−2	−1	−0.5	0	0.5	1	1.5	2	2.5	3
α	0.75	0.82	0.90	0.95	1	1.12	1.27	1.46	1.71	2.06	2.59

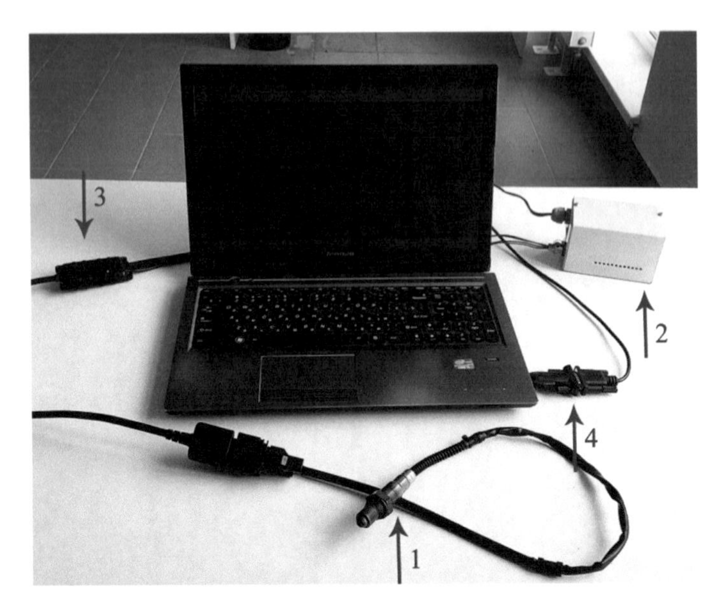

Fig. 4.14 Monitoring system for fuel combustion in boiler units: 1—broadband oxygen probe, 2—alpha indicator, 3—controller, 4—USB interface

A system for monitoring the fuel combustion in small and medium power boilers based on an oxygen probe is shown in Fig. 4.14 [23].

Technical characteristics of monitoring system for fuel combustion in boilers are given in Table 4.15.

Using a broadband oxygen sensor in the monitoring system has a number of advantages over conventional gas analyzing devices: the absence of a sampling and sample preparation system, rapid measurement of oxygen concentration (0.1–0.2 s), uninterrupted operation, long service life, easy installation for various types of thermal aggregates.

In general, the monitoring system allows:

- optimizing the fuel combustion, taking into account actual conditions, boiler operating conditions and fuel characteristics;
- reducing fuel consumption for at least 10%;
- reducing nitrogen oxide emissions up to 40%;
- reducing the level of carbon monoxide emissions up to 50%;
- increasing the boiler efficiency for at least 5%;
- simplifying staff work operations.

The features of the developed fuel combustion monitoring system in boilers of small and medium power make it possible to use it for the automatic control system of the fuel combustion process. Figure 4.15 shows a structural diagram of such system functioning.

Table 4.15 Technical characteristics of the fuel combustion monitoring system

Parameter	Value
Output signal of the measuring probe, V	+0.1…+5.0
Recall (time delay of indication) for 50% step perturbation, s	0.1…0.3
Initial preparation time for measurements, s	≤30.0
Measuring range of the parameter α	0.5…1.5
Relative error, %	3
Indication of measurement results	LED
Cable length, m	≤5
Ambient temperature at relative humidity up to 80%	
Display unit, °C	5…50
Boxes of the measuring probe, °C	5…70
Conditions at the measurement point	
Ambient temperature, °C	50…250
Flow rate, m/s	≤15
Pressure, Pa	≤±500

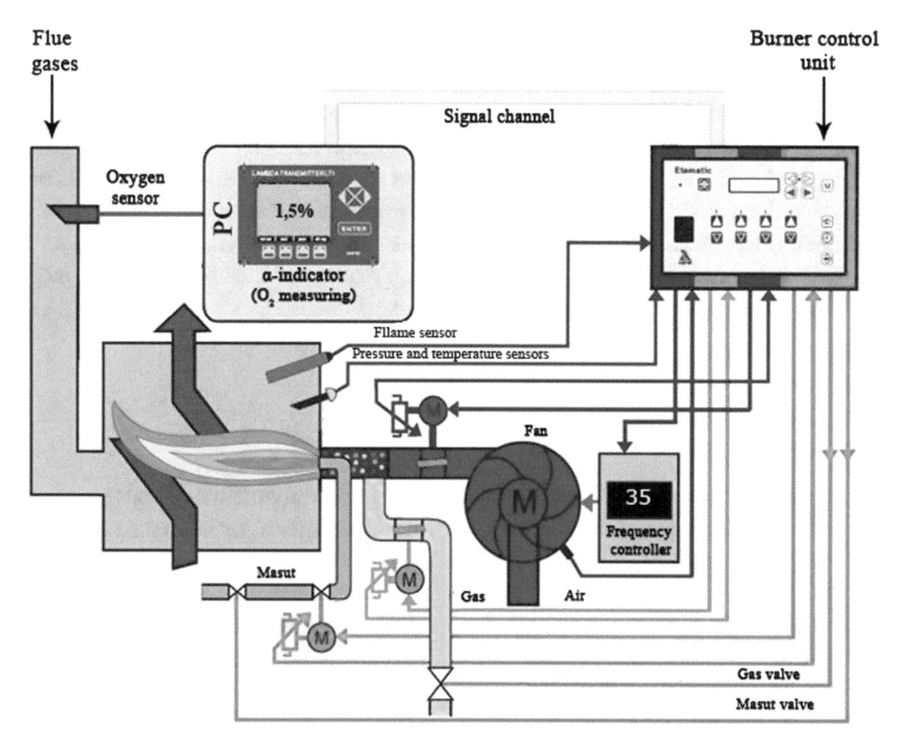

Fig. 4.15 Block diagram of automatic control system of the fuel combustion process in boiler units of small and average power on the basis of an oxygen probe

Main purpose of the control system is the fan motor speed regulation. This will encourage the optimum combustion regime in the boiler furnace, that is, provide the most favorable conditions for complete fuel combustion. The system supplies the required quantity of air to the furnace based on information received from its primary sensors (oxygen probe, temperature sensors and vacuum).

The task of maintaining the optimum combustion regime is ensured by selecting the necessary speed of motors rotation of the traction mechanisms with fully open guide devices in practically the entire range of the operating capacity of the O_2-corrected combustion.

The specified control system provides:

- rational natural gas consumption (saves 5–10% per year);
- reduction of electric power consumption by asynchronous wired motors of traction mechanisms (saves of 30–75% per year);
- reduction of emission to atmosphere due to complete fuel combustion.

4.5 Application of Developed Methods to a Renewable Energy Expert System

Policy makers over the world show increasing concern about the environment. There is considerable interest in using novel methods to generate electrical energy including wind, wave, and solar energy sources. The main reason is that conventional fuels are limited and expensive, whereas other forms, known as the renewables, are limitless and cheap. Renewable energy is expected to capture a growing percent of world energy market over the next 20 years. The key drivers of the "Renewable Energy Revolution" are [24–28]:

- Increasing global energy demand.
- Concerns about carbon emissions.
- Concerns about energy independence.
- Falling cost of renewable energy.

Despite well recognized social benefits of renewable sources of energy, the main driver for the development of these types of energy projects, or indeed any type, is their financial viability [29, 30]. Few would argue that generating electricity from the wind makes environmental sense. Wind is a clean, renewable energy source and, if sited sensitively, wind parks have limited environmental impact. The economic viability of a Wind Energy Conversion System (WECS) for a particular wind pattern depends on the success with which the system extracts the maximum possible energy in the most efficient way.

Reliability of wind generators is a very important problem. Development of wind generators diagnostic systems and their implementation to the modern wind energy generators can improve their reliability.

In today's competitive power market, cost effective equipment servicing is believed by many to be one of the most important parts of the business—for manufacturers and plant owners alike. Even financing or insuring power plant sees it as a key ingredient in running a successful generating business [31].

SmartSignal's equipment condition monitoring (eCM). Unlike preventive maintenance practices, which recommend maintenance based on failure statistics for a class of equipment problems over time, predictive condition monitoring provides equipment-specific, condition-based early warning. Predictive condition monitoring products, like SmartSignal's equipment condition monitoring (eCM) [32], provide advanced, equipment-specific warning of deteriorating conditions leading to failure or poor performance of all equipment makes and types.

In the power generation industry, predictive condition monitoring provides early warning of failure of combustion turbines, steam turbines, boiler feedwater pumps, coal pulverizers, electrostatic precipitators and cooling water pumps.

Compared to traditional technologies, SmartSignal eCM demonstrates significant advantages such as monitoring multiple equipment operating modes like partial load conditions, analyzing multiple equipment, and creating serial number specific models.

Comparing SmartSignal eCM to traditional threshold limit technology helps explain the predictive technology. In traditional monitoring, the manufacturer recommended upper and lower sensor threshold limits to initiate machine trips to avoid damage. Manufacturers set the thresholds based on deep first principle understanding of the equipment design parameters. In contrast, the SmartSignal eCM models sensor values of normal equipment performance instead of design parameters for each serial number piece of monitored equipment. Doing so enables the software to quickly deploy enterprise wide fleet monitoring solutions compared to other technology such as neural network applications.

To start up, the SmartSignal eCM uses plant historical data to create a personalized, empirical model of the equipment's normal operation range. Both "personalized" and "empirical" are key distinction: personalized because the model is for that equipment serial number and empirical because the model is built using only the actual operating data, no detailed engineering knowledge of the engine is needed on the stage. This empirical serial number specific model of normal equipment operation creates an estimate for each sensor, in real-time, based on values of correlated sensors. The software compares actual sensor values to estimated sensor values and detects subtle but significant differences, called residuals (residuals provide the basis for early warning of abnormal equipment conditions.

From the process standpoint, eCM starts by collecting a "snapshot" of sensor values that make up an eCM model. eCM automatically creates an empirical model of normal performance of the asset using that statistical "snapshot". Unlike other monitoring techniques, in real-time the eCM empirical model generates an estimated value for each sensor that would be characteristic of normal operation. As previously noted, each sensor estimate is based not only on that sensor's history but also based on how that sensor interacted with every other sensor value.

The result is a data-driven empirical model of each asset. Then, in real-time, the software effectively removes the effect of normal operation by subtracting in estimated values from the actual values just collected to generate "residuals" . If the equipment is running normally, the resulting residuals should be small and evenly distributed around zero.

Equipment faults show up as spikes or trends in the residuals. eCM compared the residuals using a patented statistical technique. If significant deviations are found the equipment is running abnormally, and the eCM issues an "alert" . The "alert" are fed into diagnostic rules engine, which analyses the pattern of alerts to see if this is a known pattern or if it meets the preestablished criteria for being promote into an item on the WatchList and/or notifying an analyst (these criteria are set-up during the SmartStart Installation methodology phase).

Information about the incident is fed back into the eCM database and all the information can be sent back to the control system.

Many diagnostic professionals have relatively little knowledge about the processes they have to analyze, and sometimes they even try to tune existing diagnostic (or expert) systems or develop new diagnosis strategies without sufficient process knowledge [33]. However, analyzed process knowledge is one, if not the most important, factor in achieving effective control. The better we know the behavior and specific of the process, the better able are we to choose the right diagnostic scheme and to find the best suited tuning for the given situation and performance requirement.

To achieve good diagnostic performance we must have knowledge regarding:

- the process;
- process diagnostic;
- the diagnostic system.

Knowledge of the process, is quite likely the most important of the three. Included in this process knowledge should be the type of the process behavior we have to deal with as well as the static and dynamic process parameters that are involved.

Knowledge of process diagnostic, is the most obvious. Without a thorough understanding of the fundamental behavior of diagnostic systems, and the various approaches used to configure and tune them, we can newer (or least not in an acceptable amount of time) achieve reasonable diagnostic performance.

Diagnostic system, is fairly easy to understand. Without an in-depth knowledge of the distributed diagnostic system and its features, we are in no position to even attempt to implement diagnostic scheme.

General information about the process is certainly needed, but is not sufficient to achieve optimum diagnostics.

Ukrainian complex wind generators building program was started in 1997. Implementation of wind generators to Ukrainian power systems is an important issue. Institute of renewable energy engineering was founded in the structure of the Ukrainian National Academy of Sciences in 2003. One of Institute scientific areas is development of wind energy systems. Modern wind energy system is a complex constructive building. One of the main peculiarity of wind generator operation is irregular load of

Fig. 4.16 Photo of the expert system prototype

it's construction and rotating parts, particularly given the inability of wind to generate power consistently. Dynamic loads to different parts of wind generator and wind farm can be significant. Special monitoring systems and diagnostic systems should be implemented to modern wind energy system.

A prototype of an expert system for wind energy systems vibration diagnostics was developed at the Institute of Electrodynamics of the Ukrainian National Academy of Sciences (Fig. 4.16).

A stochastic approach was used for the expert system software elaboration. Some of the methods were demonstrated in [34]. Vibration signals of wind energy system parts are considered as stochastic processes. Early warning of failures (defects) of the Wind Energy Conversion System (WECS) parts is based on the changes of statistical parameters and characters of vibration processes which accompanies wind energy system operation.

A prototype of the expert system provides vibration measurement of the following WECS elements:

- wind turbine bearings;
- generator bearings;
- transmission;
- wind farm.

The prototype of expert system also has possibilities:

- to measure aerodynamic noise during wind generation operation;
- to accumulate and record measured data;
- to analyze the measured data.

The central module for measured data registration and transmission is constructed for WECS on the base of PC. It is located directly in wind generator farm of WECS. Module for measured data input/output and storage is located on central module. Measured data from every wind generator of WECS comes to central module for further data registration and transmission. Radio channels, cable communications, WEB modems can be used for information transmission to the central module.

Special devices, on PC base, were constructed and manufactured for WECS information data measurement. It makes possible to provide a preliminary analysis of input signals, analog-to-digital (A-D) conversion of input signals, data registration and storage. ADC module converts amplified analog information signals to digital signals. Sampling frequencies are the following: 62.5, 125, 250, 500 kHz, 1, 2 MHz. 12 bit ADC and buffer storage 128 kBit embedded to the prototype of the expert system.

Software description. Original software includes four modules:

1. Basic software module.
2. Module of input information signals storage and analysis.
3. Module of output information visualization.
4. Interface module for analog data input.

Basic software module connects and controls all others modules.

Figure 4.17 shows software structure and Fig. 4.18 shows basic software windows (plug-ins). The software provides "fast search" by predetermined parameters that provides fast search of the necessary information. Special "map" is formed for

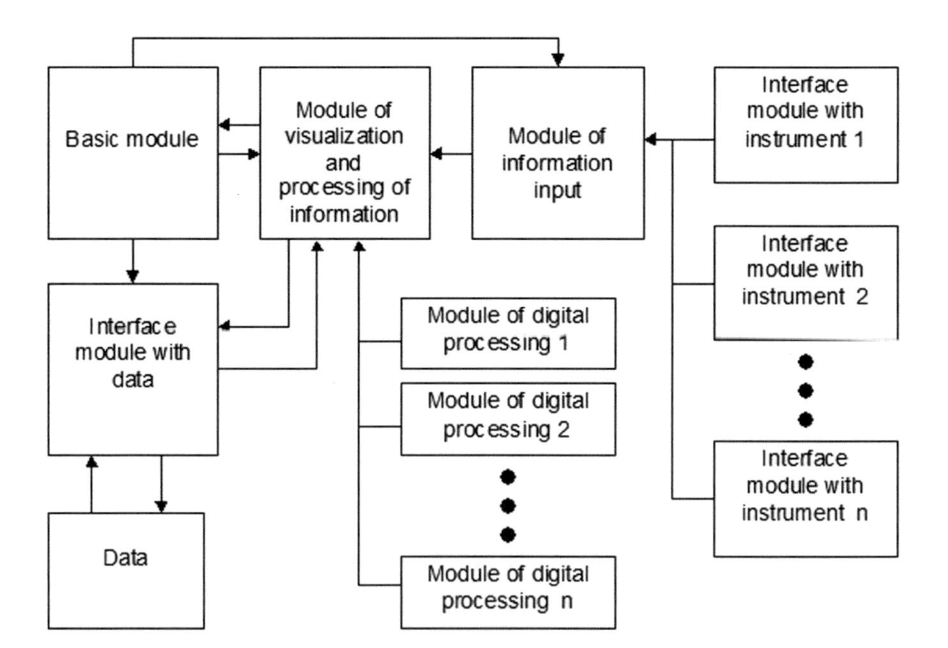

Fig. 4.17 Structure of the developed software

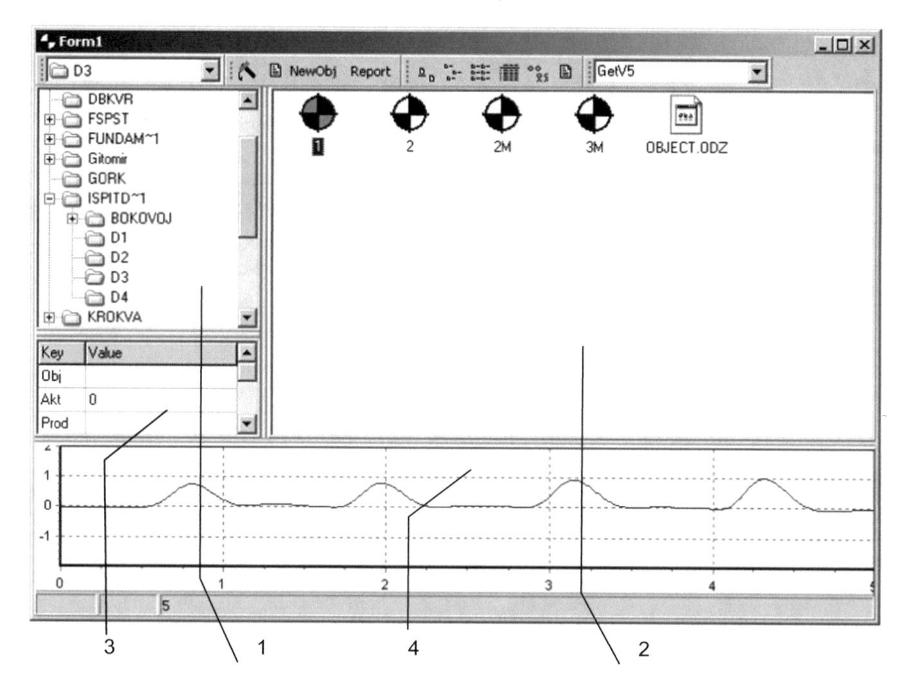

Fig. 4.18 Basic software program window: 1—structure; 2—test list; 3—testing "map"; 4—preliminary data scan

every measurement (location of wind system, time of experiment, etc.). Window "Visualization and analysis" is used for data scan and for new data records (Fig. 4.19).

The software interface is based on the "Dock Management" technology. It permits to develop a flexible software interface. Data analysis modules are dynamic libraries that include analysis procedures and visual element for determination of parameters.

Software includes the following signal processing methods:

- histogram analysis of signals and data with Pearson's smoothing curves;
- autocorrelation analysis of information signals;
- statistical spectral analysis;
- autoregressive analysis.

The software also has a decision-making module. Some of the methods are represented in [35].

It is necessary to remark that input data is not changed during analysis and program recording of files, used in algorithms and their parameters.

If it is necessary data may be converted to the format "XY" and others software packages may be applied for information signal processing and data analysis.

Special "Mark" is realized in the developed software interface (Fig. 4.20).

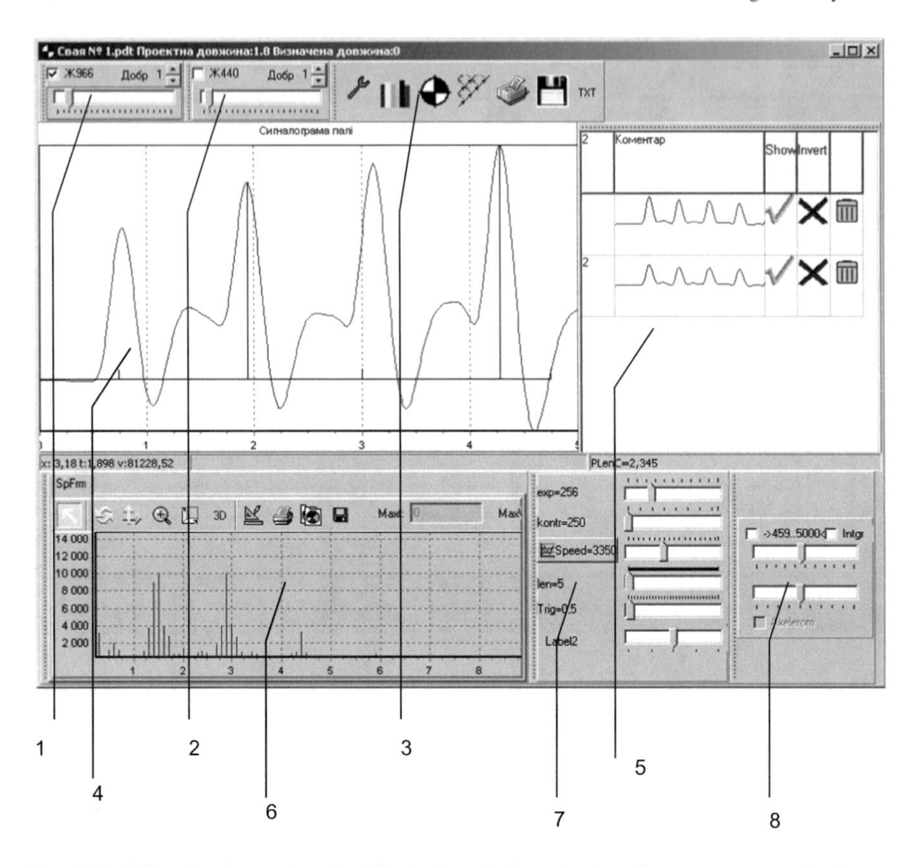

Fig. 4.19 "Visualization and analysis" window: 1, 2—rejection filters parameters; 3—instruments panel; 4—signal processing; 5—realizations list; 6—process spectrum; 7—visualization information parameters; 8—band-pass filters parameters

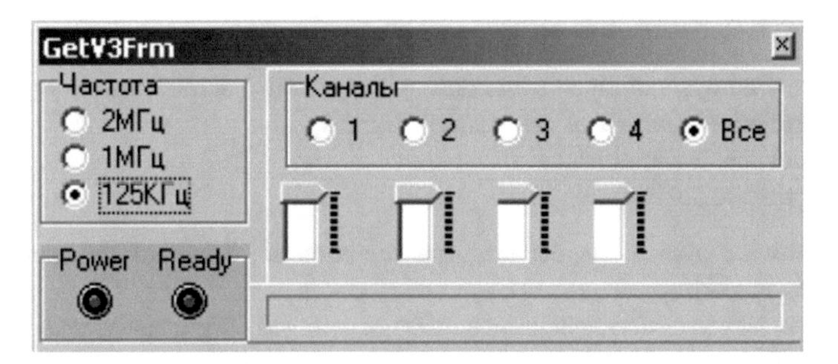

Fig. 4.20 "Mark" dialog window

It allows connecting others software packages without basic software program code changes. It is necessary to remark that that window “Mark” is reflected together with window “Visualization and analysis”. “Mark” is a part of the window.

Some expert system prototype specifications are the following:

Physical signals, measuring by the expert system prototype:

- Vibration signals (vibration displacement, vibration velocity, vibration acceleration)
- Technical specification of the expert system prototype:

Frequency range of measured vibration signals: 20 Hz–30 kHz.

- Range of analog signals sampling frequencies: 16 kHz–1 MHz.
- Maximal duration of vibration signals 1 s.
- Amplitude range of input vibration signals ± 1.024 V.
- Number of measurement channels: 4.

The prototype of expert system was used for studying the vibration parameters of USW 56–100 wind turbine at the Ukrainian corporation “PO Yuzhmash” . Histogram analysis with Pearson smoothing curves application and statistical spectral analysis was performed during the experimental study. Different types of vibration transducers were used during the research.

Experimental study was performed in so called “engine” regime of wind turbine. Generator’s shaft was rotating at the average speed of 1449 revolution per minute (rev./min). Wind wheel hub shift was rotating at the speed of 72 rev/min.

The following transducers were installed at the wind turbine: Vp1 (accelerometer DN-4), VP2 (accelerometer D-14), VP3 (accelerometer ABC-017), VP4 (seismic transducer SV-10C). Different types of transducers were used because the most effective transducer type for WECS vibrodiagnostics should be selected.

Transducers VP1-VP3 were installed at different parts of wind turbine. Transducers installation places are shown in Fig. 4.21.

Sampling frequency equals 15 625 Hz during test. Volume of studied vibration signals samples was N = 16 000. Estimations of spectral and correlation functions and distribution functions were calculated.

Estimation of normalized autocorrelation function of vibration signals from transducer VP1 which was installed at the frame of transmission near bearing of low rotating speed shift in radial direction is shown in Fig. 4.22. Estimation of power spectral density of vibration signals from transducer VP1 which was installed at the frame of transmission near bearing of low rotating speed shift in radial direction is shown in Fig. 4.23.

Histogram with Smoothing curve of vibration signals from transducer VP1 which was installed at the frame of transmission near bearing of low rotating speed shift in radial direction is shown in Fig. 4.24.

Using the developed prototype of WECS expert system diagnostic parameters such as autoregression parameters were estimated. The estimations of diagnostic parameters were analyzed. Obtained results shows that diagnostic algorithms and software can be applied for the WECS parts vibration diagnostics.

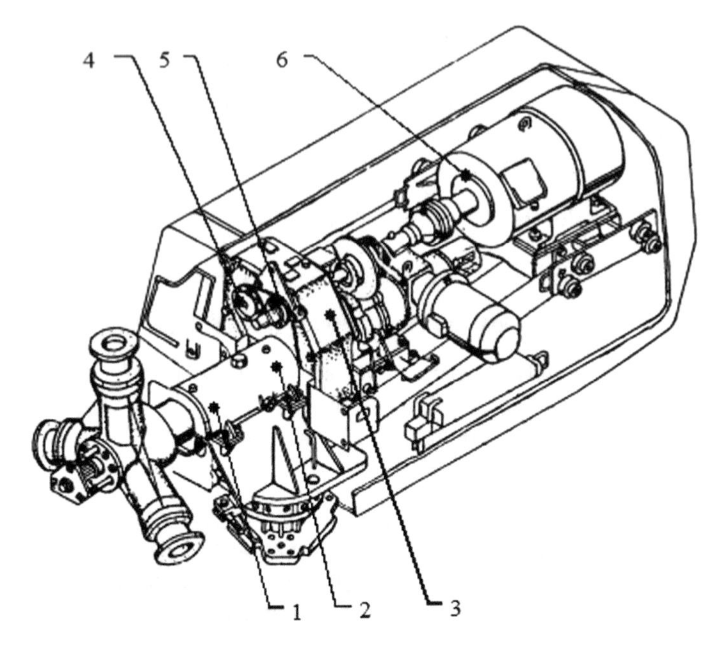

Fig. 4.21 Places of transducers installation: 1—frame of basic shift in radial direction near bearing at the side of wind wheel hub; 2—frame of basic shift in radial direction near bearing at the side of transmission; 3—frame of transmission near bearing of low rotating speed shift in radial direction; 4—frame of transmission near bearing of middle rotating speed shift in axis direction; 5—frame of transmission near bearing of high rotating speed shift in axis direction; 6—frame of generator near working end of shift in radial direction

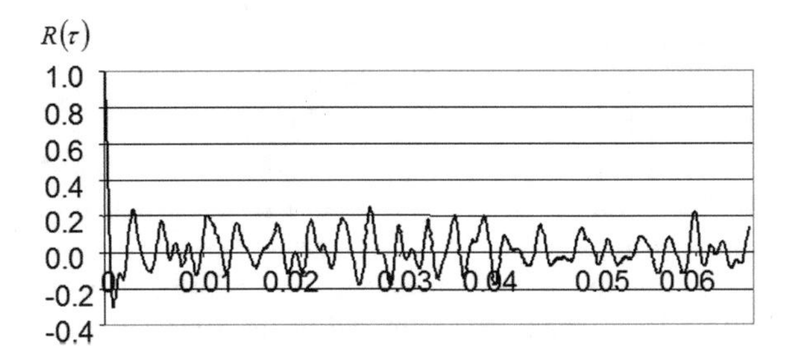

Fig. 4.22 Estimation of autocorrelation function of vibration signal at the frame of transmission

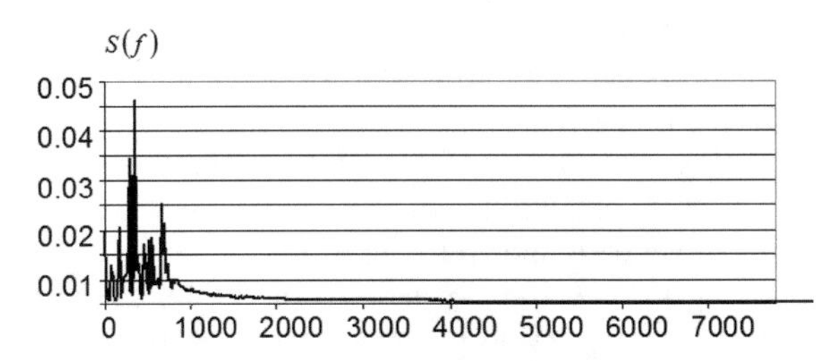

Fig. 4.23 Estimation of power spectral density function of vibration signal at the frame of transmission

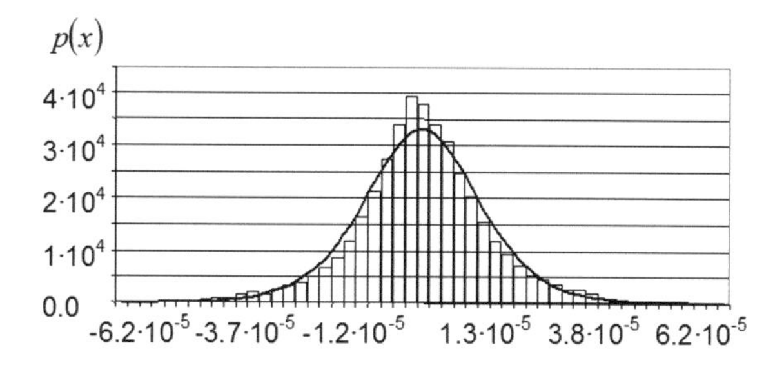

Fig. 4.24 Histogram and smoothing curve of Pearson type 7 of vibration signal at the frame of transmission

References

1. Uomoto, T.: Non-Destructive Testing in Civil Engineering 2000 (2000). ISBN: 9780080545356
2. Myslovych, M., Sysak, R.: Design peculiarities of multi-level systems for technical diagnostics of electrical machines. Comput. Prob. Electr. Eng. **1**(4), 47–50 (2014)
3. Dmitriev, S.A., Manusov, V.Z., Ahyoev, J.S.: Diagnosing of the current technical condition of electric equipment on the basis of expert models with fuzzy logic. In: 2016 57th International Scientific Conference on Power and Electrical Engineering of Riga Technical University (RTUCON), Riga, Latvia, 13–14 Oct 2016. https://doi.org/10.1109/rtucon.2016.7763126
4. Kinney, P.: Zigbee technology: wireless control that simply works. Commun. Des. Conf. **2**, 1–7 (2003)
5. Blevins, T., Chen, D., Nixon, M., Wojsznis, W.: Wireless Control Foundation: Continuous and Discrete Control for the Process Industry (2015). ISBN: 978-0-876640-88-3
6. Jo, M., Maksymyuk, T., Batista, R.L., Maciel, T.F., de Almeida, A.L.F., Klymash, M.: A survey of converging solutions for heterogeneous mobile networks. IEEE Wirel. Commun. **21**(6), 54–62 (2014). https://doi.org/10.1109/MWC.2014.7000972
7. Yang, J., Zhou, J., Lv, Z., Wei, W., Song, H.: A real-time monitoring system of industry carbon monoxide based on wireless sensor networks. Sensors **15**(11), 29535–29546 (2015). https://doi.org/10.3390/s151129535
8. Fang, H., Xia, J., Zhu, K., Su, Y., Jiang, Y.: Industrial waste heat utilization for low temperature district heating. Energy Policy **62**, 236–246 (2013). https://doi.org/10.1016/j.enpol.2013.06.104
9. Allan, R.N., Billinton, R.: Reliability Evaluation of Power Systems (1996). ISBN: 978-1-4899-1860-4
10. Fan, Z., Kulkarni, P., Gormus, S., Efthymiou, C., Kalogridis, G., Sooriyabandara, M., Zhu, Z., Lambotharan, S., Chin, W.H.: Smart grid communications: overview of research challenges, solutions, and standardization activities. IEEE Commun. Surv. Tutor. **15**(1), 21–38 (2013). https://doi.org/10.1109/SURV.2011.122211.00021
11. Dekusha, L., Kovtun, S., Dekusha, O.: Heat flux control in non-stationary conditions for industry applications. In: 2019 IEEE 2nd Ukraine Conference on Electrical and Computer Engineering, Lviv, Ukraine, 2–6 July 2019, pp. 601–505. https://doi.org/10.1109/ukrcon.2019.8879847
12. Babak, V., Dekusha, O., Kovtun, S., Ivanov, S.: Information-measuring system for monitoring thermal resistance. In: CEUR Workshop Proceedings, vol. 2387, pp. 102–110 (2019). http://ceur-ws.org/Vol-2387/20190102.pdf
13. Dekusha, O., Babak, V., Vorobiov, L., Dekusha, L., Kobzar, S., Ivanov, S.: The heat exchange simulation in the device for measuring the emissivity of coatings and material surfaces. In: IEEE 39th International Conference on Electronics and Nanotechnology (ELNANO), 16–18 April 2019, Kyiv, Ukraine. p. 301–304. https://doi.org/10.1109/elnano.2019.8783537
14. Lee, J., Wu, F., Zhao, W., Ghaffari, M., Liao, L., Siegel, D.: Prognostics and health management design for rotary machinery systems—reviews, methodology and applications. Mech. Syst. Signal Process. **1–2**(42), 314–334 (2014). https://doi.org/10.1016/j.ymssp.2013.06.004

15. Zhenhua, W., Xiaojun, M., Hongfu, Z.: Characteristics analysis and experiment verification of electrostatic sensor for aero-engine exhaust gas monitoring. Measurement **47**, 633–644 (2014). https://doi.org/10.1016/j.measurement.2013.09.041
16. Dubovikov, O.A., Brichkin, V.N., Loginov, D.A.: Study of the possible use of producer gas of coal gasification as fuel. In: Litvinenko V. (eds) XVIII International Coal Preparation Congress, pp. 593–599 (2016). https://doi.org/10.1007/978-3-319-40943-6_91
17. Wei, L., Geng, P.: A review on natural gas/diesel dual fuel combustion, emissions and performance. Fuel Process. Technol. **142**, 264–278 (2016). https://doi.org/10.1016/j.fuproc.2015.09.018
18. Majid, Z.A., Mohsin, R., Shihnan, A.H.: Effect of biodiesel blends on engine performance and exhaust emission for diesel dual fuel engine. Energy Convers. Manag. **88**, 821–828 (2014). https://doi.org/10.11113/jt.v78.3199
19. Schnick, M., Dreher, M., Zschetzsche, J., Fussel, U., Spille-Kohoff, A.: Visualization and optimization of shielding gas flows in arc welding. Weld. World **1–2**(56), 54–61 (2012). https://doi.org/10.1007/BF03321146
20. Isles, J.: Servicing for the long term. Power Eng. Int. **11**(10), 36–40 (2003)
21. Hotlan, T.P.: Early warning system. Power Eng. Int. **11**(9), 39–43 (2003)
22. Brockwell, P.J., Linder, A.: Prediction of Lévy-driven CARMA processes. J. Econ. **2**(189), 263–271 (2015). https://doi.org/10.1016/j.jeconom.2015.03.021
23. Appadoo, S.S., Thavaneswaran, A., Mandal, S.: RCA model with quadratic GARCH innovation distribution. Appl. Math. Lett. **25**(10), 1452–1457 (2012). https://doi.org/10.1016/j.aml.2011.12.023
24. Popov, O., Iatsyshyn, A., Kovach, V., Artemchuk, V., Taraduda, D., Sobyna, V., Sokolov, D., Dement, M., Yatsyshyn, T., Matvieieva, I.: Analysis of possible causes of NPP emergencies to minimize risk of their occurrence. Nucl. Radiat. Saf. **1**(81), 75–80 (2019). https://doi.org/10.32918/nrs.2019.1(81).13
25. Popov, O., Iatsyshyn, A., Kovach, V., Artemchuk, V., Taraduda, D., Sobyna, V., Sokolov, D., Dement, M., Yatsyshyn, T.: Conceptual approaches for development of informatioinal and analytical expert system for assessing the npp impact on the environment. Nucl. Radiat. Saf. **3**(79), 56–65 (2019). https://doi.org/10.32918/nrs.2018.3(79).09
26. Shkitsa, L.E., Yatsyshyn, T.M., Popov, A.A., Artemchuk, V.A.: The development of mathematical tools for ecological safe of atmosphere on the drilling well area. Oil Ind. **11**, 136–140
27. Yatsyshyn, T., Shkitsa, L., Popov, O., Liakh, M. Development of mathematical models of gas leakage and its propagation in atmospheric air at an emergency gas well gushing. Eastern-Eur. J. Enterp. Technol. **5**, **10**(101), 49–59 (2019). https://doi.org/10.15587/1729-4061.2019.179097
28. Barlas, T.K., van Kuik, G.A.M.: Review of state of the art in smart rotor control research for wind turbines. Prog. Aerosp. Sci. **46**(1), 1–27 (2010). https://doi.org/10.1016/j.paerosci.2009.08.002
29. Verdejo, H., Escudero, W., Kliemann, W., Awerkin, A., Becker, C., Vargas, L.: Impact of wind power generation on a large scale power system using stochastic linear stability. Appl. Math. Model. **40**(17–18), 7977–7987 (2016). https://doi.org/10.1016/j.apm.2016.04.020
30. Zaporozhets, A.: Analysis of control system of fuel combustion in boilers with oxygen sensor. Periodica Polytech. Mech. Eng. **64**(4), 241–248 (2019). https://doi.org/10.3311/PPme.12572
31. Zaporozhets, A., Kovtun, S., Dekusha, O.: System for montoring the technical state of heating networks based on UAVs. In: Shakhovska, N., Medykovskyy, M. (eds.) Advances in Intelligent Systems and Computing IV, vol. 1080. Springer, Cham, pp. 935–950 (2020). https://doi.org/10.1007/978-3-030-33695-0_61
32. Babak, V.P., Mokiychuk, V., Zaporozhets, A., Redko, O.: Improving the efficiency of fuel combustion with regard to the uncertainty of measuring oxygen concentration. Eastern-Eur. J. Enterp. Technol. **6, 8**(84), 54–59 (2016). https://doi.org/10.15587/1729-4061.2016.85408
33. Zaporozhets, A.O., Redko, O.O., Babak, V.P., Eremenko, V.S., Mokiychuk, V.M.: Method of indirect measurement of oxygen concentration in the air. Naukovyi Visnyk Natsionalnoho Hirnychoho Universytetu **5**, 105–114 (2018). https://doi.org/10.29202/nvngu/2018-5/14

34. Zaporozhets, A.: Development of software for fuel combustion control system based on frequency regulator. In: CEUR Workshop Proceedings, vol. 2387, pp. 223–230 (2019). http://ceur-ws.org/Vol-2387/20190223.pdf
35. Zimroz, R., Bartelmus, W., Barszcz, T., Urbanek, J.: Diagnostics of bearings in presence of strong operating conditions non-stationarity—a procedure of load-dependent features processing with application to wind turbine bearings. Mech. Syst. Signal Process. **46**(1), 16–27 (2014). https://doi.org/10.1016/j.ymssp.2013.09.010

Printed in the United States
by Baker & Taylor Publisher Services